Cash Incentives and Military Enlistment, Attrition, and Reenlistment

Beth J. Asch, Paul Heaton, James Hosek, Francisco Martorell, Curtis Simon, John T. Warner

Prepared for the Office of the Secretary of Defense

NATIONAL DEFENSE RESEARCH INSTITUTE

The research described in this report was prepared for the Office of the Secretary of Defense (OSD). The research was conducted in the RAND National Defense Research Institute, a federally funded research and development center sponsored by the OSD, the Joint Staff, the Unified Combatant Commands, the Department of the Navy, the Marine Corps, the defense agencies, and the defense Intelligence Community under Contract W74V8H-06-C-0002.

Library of Congress Control Number: 2010927929

ISBN: 978-0-8330-4966-7

The RAND Corporation is a nonprofit research organization providing objective analysis and effective solutions that address the challenges facing the public and private sectors around the world. RAND's publications do not necessarily reflect the opinions of its research clients and sponsors.

RAND® is a registered trademark.

Published 2010 by the RAND Corporation
1776 Main Street, P.O. Box 2138, Santa Monica, CA 90407-2138
1200 South Hayes Street, Arlington, VA 22202-5050
4570 Fifth Avenue, Suite 600, Pittsburgh, PA 15213-2665
RAND URL: http://www.rand.org/
To order RAND documents or to obtain additional information, contact
Distribution Services: Telephone: (310) 451-7002;
Fax: (310) 451-6915; Email: order@rand.org

Preface

Between fiscal year (FY) 2000 and FY 2008, the real Department of Defense (DoD) budget for enlistment and reenlistment bonuses increased substantially, from $266 million to $625 million (in FY 2008 dollars) for enlistment bonuses and from $891 million to $1.4 billion for selective reenlistment bonuses (Department of Defense, various years). Bonus increases were a response to rising manpower requirements in the case of the Army and Marine Corps, declines in youth attitudes toward the military as the Iraq War unfolded, and increases in the frequency and duration of hazardous deployments. Congress and the Government Accountability Office have raised questions about the effectiveness of bonuses, what the services received for this large increase in bonuses, whether bonuses were paid to individuals who would have enlisted or reenlisted in the absence of bonuses, and whether other policies might have been more effective in maintaining or increasing the supply of personnel to the armed forces.

This monograph provides an empirical analysis of the enlistment, attrition, and reenlistment effects of bonuses, applying statistical models that control for such other factors as recruiting resources, in the case of enlistment and deployments in the case of reenlistment, and demographics. Enlistment and attrition models are estimated for the Army and our reenlistment model approach is twofold. The Army has greatly increased its use of reenlistment bonuses since FY 2004, and we begin by providing an in-depth history of the many changes in its reenlistment bonus program during this decade. We follow this with two independent analyses of the effect of bonuses on Army reenlistment. As we show, the results from the models are consistent, lending credence to the robustness of the estimates. One approach is extended to the Navy, the Marine Corps, and the Air Force, to obtain estimates of the effect of bonuses on reenlistment for all services. We also estimate an enlistment model for the Navy. The estimated models are used to address questions about the cost-effectiveness of bonuses and their effects in offsetting other factors that might adversely affect recruiting and retention, such as changes in the civilian economy and frequent deployments. The report should be of interest to policymakers concerned with military recruiting and retention and to defense manpower researchers.

This research was sponsored by the Office of Accession Policy within the Office of the Under Secretary of Defense for Personnel and Readiness and was conducted within the Forces and Resources Policy Center of the RAND National Defense Research Institute, a federally funded research and development center sponsored by the Office of the Secretary of Defense, the Joint Staff, the Unified Combatant Commands, the Navy, the Marine Corps, the defense agencies, and the defense Intelligence Community.

For more information on RAND's Forces and Resources Policy Center, contact the Director, James Hosek, by email at James_Hosek@rand.org; by phone at 310-393-0411, extension 7183; or by mail at the RAND Corporation, 1776 Main Street, P.O. Box 2138, Santa Monica, California 90407-2138. More information about RAND is available at http://www.rand.org.

Contents

Figures

Tables

Summary

Until recently, the wars in Afghanistan and Iraq placed great stress on military recruiting and retention. Recruit quality fell between FY 2003 and FY 2008 while the services, particularly the Army, struggled to meet its overall recruiting goal. Recruit quality refers to recruits who are high school graduates and who score in the top half of the Armed Forces Qualification Test (AFQT). Between FY 2004 and FY 2008, the percentage of active duty, non-prior service, high school graduate recruits fell from 95 percent to 92 percent, for all of DoD, and the percentage with AFQT scores above 50 fell from 73 percent to 68 percent (Department of Defense, 2009). In FY 2005, the Army failed to meet its overall recruiting goal. As for retention, first- and second-term reenlistment rates remained fairly stable over the period FY 1996 to FY 2007 (Hosek and Martorell, 2009). But, as the burden of deployment on individual soldiers increased under long and multiple deployments, the effect of deployments on reenlistment, which had been positive, became negative in FY 2006 and FY 2007. Also, the Army imposed stop-loss on a significant fraction of its enlisted force beginning in FY 2003.[1] The recent recession has since helped recruiting and retention; for example, in FY 2009, the percentage of recruits with a high school diploma increased to 96 percent, and the percentage with AFQT scores above 50 increased to 72 percent.

To address the recruiting and retention challenges, budgets for enlistment and reenlistment bonuses increased dramatically beginning in FY 2004. In FY 2008 dollars, the selective reenlistment bonus (SRB) budget for active duty personnel increased across DoD from $625 million in FY 2003 to $1.4 billion in FY 2008. The bulk of the increase was due to increases in the Army and Marine Corps budgets. In FY 2003, these budgets were 21 percent of the DoD SRB budget. By FY 2008, these budgets were 66 percent of the DoD SRB budget. The large ramp-up in enlistment bonuses began in FY 2006. In FY 2005, the DoD enlistment bonus (EB) budget increased from $296 million (in FY 2008 dollars) to $475 million in FY 2006, ultimately reaching $611 million in FY 2008. These increases helped the services reach their recruiting and retention goals during operations in Iraq and Afghanistan and at a time when the Army and Marine Corps were increasing their end strength.

Rising bonuses in recent years have raised questions about the size, scope, and efficacy of these expenditures. Congress directed DoD to provide information on the number and average amounts of bonuses, by occupational area, and on metrics of performance. The Office of Accession Policy and the Office of Officer and Enlisted Personnel Management, within the Office of the Under Secretary of Defense for Personnel and Readiness, requested that RAND provide input to enable these offices to respond to the Congressional mandate. This report summarizes the findings of the RAND study.

---

[1] Stop-loss policies were designed to prevent soldiers whose contracts were expiring from separating from the Army.

Approach

To provide input on the size and scope of bonuses, we used data provided by the Defense Manpower Data Center (DMDC) and enlistment data provided by the Navy and the Army to compute the number of enlistment and selective reenlistment bonuses paid and their average amount, by occupation. To assess the effectiveness of bonuses, we drew from the 7th, 9th, and 10th Quadrennial Reviews of Military Compensation, together with the report of the Defense Advisory Committee on Military Compensation, to identify the criteria for assessing military compensation and, specifically, bonuses. These criteria are that bonuses (1) support DoD's force management goals, particularly recruiting and retention goals, (2) are flexibly used and can adjust quickly as circumstances change and address specific recruiting or retention problem areas, and (3) are efficient in achieving force management goals at the least cost.

We then estimated enlistment and reenlistment models to assess the extent to which bonuses contributed to recruiting and retention success (criterion 1). In the case of enlistment, we use data provided by the Army and Navy from FY 1998 to FY 2008 on enlistments and recruiting resources, including enlistment bonuses and recruiters. We also used Current Population Survey data on the civilian unemployment rate, civilian earnings, and civilian demographic characteristics. We aggregated the data by state and quarter to estimate models of the relationship between high-quality enlistments and enlistment bonuses, recruiters, military pay, variables representing civilian opportunities (such as the civilian unemployment rate), variables representing the Iraq war (e.g., casualties), and demographic characteristics of each state over time that may reflect individuals' taste for military service. Models are estimated for the Army and the Navy.

We estimate the market-expansion effects of bonuses and the effects of enlistment bonuses on first-term attrition in the Army using Army data from FY 1998 to FY 2008. This analysis provides insight into whether enlistment bonuses increase or decrease the number of person-years provided by an Army recruit during the first term.

We took advantage of two research efforts that deal with reenlistment—studies that began before the start of this project. One effort focuses on the Army and the other focuses on all four services. Each had taken steps toward developing its own database, and the Army analysis in each effort provided an opportunity to compare results to see if they were consistent and, in that sense, robust to the different methods used in the two efforts. Both efforts estimate two aspects of the effect of bonuses on reenlistment: the effect on the probability of reenlistment and the effect on the length of reenlistment (given reenlistment).

The Army-only analysis provides a detailed history of the Army reenlistment bonus program and estimates models for Zone A reenlistment (at two to six years of service) and Zone B reenlistment (seven to ten years of service).[2] The data were provided by the Defense Manpower Data Center and cover non-prior service personnel who entered the Army between FY 1988 and FY 2002 and faced a reenlistment decision between FY 2002 and FY 2006 in one of 24 Army military occupational specialties (MOSs). These 24 MOSs account for nearly half of all personnel who entered the Army between FY 1988 and FY 2002. The models control for other key variables such as length of deployment and stop-loss. The models estimated are a probit model of reenlistment defined over the period of 12 months before the expiration of term of

---

[2]   Zones A and B refer to the first and second reenlistment decision points.

service (ETS) up to the ETS, an annual model of reenlistment, and a Tobit model of the length of reenlistment.

The second effort is an analysis of first- and second-term reenlistment in each military service across all of their occupations. This analysis builds on data provided by DMDC and used for an analysis of the effect of deployment on reenlistment during the global war on terrorism presented in Hosek and Martorell (2009). We extend that study by refining the SRB measure used. The models estimated are a linear probability model of reenlistment and a Tobit model of the length of reenlistment. Both the Army-only effort and the effort that builds on Hosek and Martorell use two alternative definitions of the bonus variable. We describe this in detail in the report. A key point is that we expect the bonus effect estimates obtained for the Army from the analysis that builds on Hosek and Martorell to bracket those obtained in the Army-only analysis—and they do.

To assess the extent to which bonuses are used flexibly (criterion 2), we compare average enlistment and reenlistment bonus amounts by occupation, length of enlistment (or reenlistment), and over time. An additional way to assess flexibility is to estimate the skill-channeling effects of enlistment bonuses, or the extent to which they induce recruits to select hard-to-fill occupations. We do not estimate the skill-channeling effects of enlistment bonuses in this study. However, comparisons of bonus amounts provide information on the extent to which the services varied the incentives for enlistment and reenlistment.

To assess efficiency (criterion 3), we compare the cost of recruiting or retaining additional personnel using bonuses to the cost of using pay and other resources. It is important to note that we use a relative metric, not an absolute metric, of cost-effectiveness. Thus, our analysis does not answer the question of whether bonuses were set at the right levels. Instead, we answer the question of whether the services increased enlistments and reenlistments at a lower cost by using bonuses rather than by using other resources such as pay. More specifically, we estimate the cost per additional high-quality Army and Navy recruit and the cost per additional year of reenlistment for each service using bonuses.

Caveats

This study provides estimates of the cost-effectiveness of bonuses relative to other resources that could be used to increase enlistments and reenlistments, notably pay. The study does not estimate the absolute cost of bonuses, nor does it determine whether bonus levels were optimal and could have achieved the same effects at even lower cost. Nonetheless, the estimates presented here provide policymakers with information on whether a given expenditure will produce a larger effect if spent on bonuses or other resources. As discussed under the topic of future research, determining the optimal mix and levels of bonuses would require additional analysis, beyond the scope of the current study, of whether different levels and mixes of bonuses across occupations and terms of enlistment or reenlistment would results in more enlistments and reenlistments for the same cost than what was actually observed. Experimental data would be particularly well suited for performing such an analysis because the mix of bonuses could be varied randomly.

Rather than using experimental data, our analysis uses administrative data on enlistment and reenlistment bonuses paid to military applicants and reenlistees. Our approach may produce estimates of the effects of bonuses on enlistments and reenlistments that are potentially subject to both upward and downward biases. As is well known in statistics, a biased estimator is one where the expected value of the estimator does not equal the true value of the parameter being estimated. In the context of this study, our estimated effect of enlistment or reenlistment bonuses may possibly deviate from the true effect of these bonus programs. If the estimated effect overstates the true effect of bonuses, then the bias is upward and, conversely, if the estimated effect understates the true effect, the bias is downward. On the other hand, as discussed below, using administrative data also has advantages, and the alternative approach—conducting a randomized experiment—also has advantages and disadvantages.

There are several possible sources of bias in our analysis of administrative data. The first results from the possibility of reverse causality, a term commonly used in the econometrics literature. Reverse causality in the context of bonus effectiveness refers to the phonemenon whereby not only do bonuses affect enlistments and reenlistments but enlistments and reenlistments may, in reverse, also affect bonuses. Bonuses influence the willingness to enlist or reenlist, and our models seek to estimate the size of this positive effect. But because enlistments and reenlistments may affect bonuses, and in the opposite direction, reverse causality may impart a negative, downward bias. Why might enlistments and reenlistments have a negative effect on bonuses? Enlistment and reenlistment outcomes may influence the amount of bonuses set by policymakers. For example, the Army increased SRB multipliers dramatically in FY 2005–2006 over concern that retention would suffer during operations in Iraq and Afghanistan. Thus, policymakers increase bonuses when enlistments or reenlistments are down, imparting a downward bias on estimates of the incentive effect of bonuses on the willingness to enlist or reenlist.

We address the problem of reverse causality by using data across many occupations, and we estimate the models with "fixed effects" for occupation and time in the reenlistment models and for state and time in the enlistment models, to capture state, occupation, or time-specific differences in enlistment or reenlistment. Nonetheless, we recognize that the bias caused by reverse causality may still be present to the extent that it explains variations in enlistment within states over time, or variations in reenlistment within occupations over time.

Another source of potential bias is that additional factors that are not observed in our data and that are correlated with bonuses may increase enlistments in some states and in some occupations. These factors include recruiter or career counselor effort, the use of bonuses in occupations that are expanding as a service is growing (and the nonuse of bonuses in occupations that are contracting when a service is shrinking), local attitudes toward enlistment in the military, or incentives to choose specific occupations or locations. Omitting these other factors can impart an upward bias to the estimated bonus effects, offsetting to some extent the potential downward bias associated with reverse causality.

Yet another potential source of bias is related to whether the construction of the variable representing the SRB multiplier depends on deployment status and on the effects on reenlistments of stop-loss and the time of reenlistment decisions. Personnel facing a reenlistment decision may time that decision to occur while they are deployed to take advantage of higher bonuses given to those who reenlist while deployed. Further, because of stop-loss policies, individuals facing a reenlistment decision while deployed must stay in service and are not allowed to leave; however, they may reenlist. Consequently, higher bonuses may be associated with a greater chance of reenlistment, because bonuses are higher for deployed personnel and deployed personnel may be constrained only to reenlist. Because of the mechanistic relationship between reenlistment and deployment (as a result of stop-loss policies), the estimated effect of bonuses is biased upward to the extent that bonuses vary with deployment.

We address this source of bias by estimating models where SRB multipliers are defined to vary with deployment status and models where they are invariant to deployment status. Although the latter models avoid this source of bias because they do not depend on bonuses, this approach introduces measurement error into the definition of the bonus variable. Measurement error imparts a downward bias on the estimated bonus effect. On the other hand, the former approach introduces an upward bias. We therefore present estimates from both models, to show the range of estimated effects.

Because of these potential biases, the estimates presented in this report must be interpreted as associations between bonuses and enlistment or reenlistment. In other words, although we use the term "effect," the relationships we estimate are not causal but are correlations. On the other hand, we take steps to mitigate the potential biases, such as presenting a range of estimates, and use econometric methods that attempt to reduce the effects of these biases. Furthermore, using administrative data to estimate the effects of bonuses has several advantages. Administrative data are easier to collect and can be analyzed in a more timely manner. Also, they permit analysis of the effects of other variables of interest, such as deployment in the case of reenlistment and other recruiting resources in the case of enlistment. Finally, as detailed below, our estimates are quite robust to the alternative definitions we use, and they are consistent with estimates in previous literature on enlistment and reenlistment.

An alternative approach is to conduct a randomized national experiment to estimate the effects of bonuses. One advantage of this alternative approach is its ability to address the issue of reverse causality. On the other hand, this approach has drawbacks, as discussed by Moffitt (2004) and Heckman and Smith (1995), experimental results can also be subject to biases, and there are limitations to what can be learned from experiments. Ideally, both experimental and nonexperimental approaches should be used to study the effects of bonuses. This study contributes to the literature and policy debate by providing estimates based on nonexperimental approaches.

Results

Effectiveness

We find that enlistment bonuses were an important contributor to the Army's success in meeting its recruiting and retention objectives in recent years. Using our estimated Army enlistment model for simulation, we estimate that in the absence of the increase in enlistment bonuses that occurred between FY 2004 and FY 2008, the Army would have enlisted 26,700, or 20 percent, fewer high-quality enlistments, implying about 1,670 fewer enlistments per quarter over this period.

In the case of reenlistment, we use our estimated models to simulate the effects on the probability of reenlistment of eliminating the SRB program in FY 2007, a year when approximately 80 percent of Army reenlistees received bonuses. For the Army, we estimate that eliminating this program in FY 2007 would have reduced the probability of reenlistment in Zone A (at the first reenlistment point) from 39 percent to 35.3 percent, a sizable drop. Alternative models that we estimate produce even larger estimates of the effects of the SRB program for the Army. Nonetheless, the results are consistent in suggesting that bonuses were a critical tool for the Army in meeting its retention objectives in FY 2007.

We also simulate the effects for the other services as well. Eliminating the SRB program in the Marine Corps in FY 2007 would have reduced the probability of reenlistment in Zone A from 37 percent to 27 percent, also a sizable drop. The simulated effects for the Navy and Air Force are smaller, however, because the average reenlistment bonus multiplier was smaller than for the Marine Corps and Army, and their first-term reenlistment rates were higher. The Marine Corps and Army were the branches with the heaviest combat duties in Operaton Enduring Freedom (OEF) and Operation Iraqi Freedom (OIF). The reenlistment results for these services suggest that bonuses helped compensate for the heavy deployments associated with these operations.

We also find that the estimated effect of the SRB multiplier on Army reenlistment varies depending on how the SRB multiplier variable is constructed with respect to deployment. Both the Army-only model and the model building on Hosek and Martorell use two variants of the bonus variable, one that conditions the SRB multiplier on deployment status and one that does not. The models differ in how deployment status is defined, reflecting differences in the modeling approaches. The Army-only model defines deployment status by whether a member is deployed in the fiscal year he or she is facing the reenlistment decision. Thus, the bonus variable depends on whether deployment occurred during the fiscal year but the service member need not be deployed at the time of the reenlistment decision. By comparison, the four-service analysis defines deployment as of the same month as the reenlistment decision. Both analyses for the Army, regardless of how deployment status is defined, find a lower bonus effect when the SRB multiple variable is not conditioned on deployment and a higher estimate when it is conditioned on deployment. The difference in these effects is probably a result of the Army's use of stop-loss, as mentioned. However, because of the difference in how deployment status is defined, the results from the analysis based on Hosek and Martorell bracket the SRB reenlistment estimates found in the 24-MOS analysis that uses a wider window for deployment.

Estimates of the effect of the SRB multiplier on reenlistment for the other services show little difference, depending on whether the SRB multiplier is conditioned on deployment. The lack of difference probably arises from little if any use of stop-loss in these services. Thus, only the Army estimate is sensitive to how the SRB multiplier is defined.

Our study also estimated the effects of SRB multipliers on length of reenlistment (LOR). Both the Army-only analysis and the four-service analysis building on Hosek and Martorell allowed the effect of the multiplier to vary depending on its level and found positive effects of Army SRB multipliers on LOR at lower SRB multiplier levels. In the Army-only analysis, we found that the positive effects of the SRB multiplier diminish at higher multiplier levels at both the first term (Zone A) and the second term (Zone B). The four-service analysis found a positive effect of SRB multipliers on LOR at the first term for each service but a diminishing effect at higher multiplier levels at the second term for all four services. The estimated effects of the multiplier on LOR are smallest for the Air Force, especially at the end of the second term. The estimated effects for the Marine Corps are also less than they are for the Army, although larger than for the Air Force. Despite these service differences, the general result is that reenlistees are estimated to choose shorter terms at higher multiplier levels than at middle levels at the second term.

We suggest two hypotheses for the cause of the decreasing effect on LOR of higher-level SRB multipliers. First, the services place caps on bonus amounts so that at some point, choosing a longer term does not result in a higher bonus. Consequently, members have no incentive to choose longer terms and may choose shorter terms that result in the same bonus amount.

As bonuses increase, the caps are more likely to be binding. We investigated this hypothesis by estimating the SRBM effect on LOR using Army data before FY 2005 when bonus caps were higher. We find that the diminishing effects of the SRB multiplier on LOR occur at even higher multipliers when we use pre-FY 2005 data, suggesting that bonus caps did play a role for the Army.

Second, bonuses may have a diminishing effect on LOR as the multiplier increases because of an "income effect," whereby reenlistees faced with a higher multiplier choose shorter term lengths that give them the flexibility to leave earlier to take advantage of civilian opportunities. A third hypothesis may also be important. Reenlistees may have limited flexibility to choose term length, especially in some occupational areas. For example, a service might expect or constrain the service member to choose a four- or six-year reenlistment, and increases in the bonus multiplier might have little effect on the length chosen. This may be the case for the Air Force especially, which experienced the smallest bonus effect on length of reenlistment. Consequently, reenlistees may not be at liberty to increase term length when the multiplier increases.

The first two explanations suggest the possibility of improving the effectiveness of SRB multipliers. Bonus caps could be more actively managed so that increases in multipliers do not provide incentives to choose shorter term lengths. Put differently, if bonus caps are not increased when multipliers are increased, this may create a perverse incentive to choose a shorter LOR. In addition, the services may need to give members more flexibility to increase term length as the multiplier increases. However, we have not verified that service policies have limited the flexibility in choosing the length of reenlistment. The third explanation, however, is a supply side response: Service members may prefer a shorter LOR when the SRB multiplier is increased because they can still receive the same size bonus but with a shorter commitment to the service, i.e., they have more opportunity to leave sooner if they chose to do so. To summarize, an efficient bonus program should target bonuses to critical skills where the retention need is greatest and should not impede the full effect using bonus caps or limited flexibility in choosing the length of reenlistment. These are general points and there may be cases where exceptions should apply, yet the argument for exception should be explicit and well understood.

Flexibility

Comparisons of the percentage of individuals receiving bonuses and the average bonus amounts over time, across occupation, and across service length indicate that at certain times, the majority of Army recruits received an enlistment bonus and the majority of Army reenlistees received an SRB. The share of Army enlistments receiving bonuses rose from about 40 percent in September 2004 to about 70 percent in September 2008 and 80 percent of reenlistees in FY 2007 received an SRB. Thus, the Army increasingly used enlistment bonuses to expand the market and used selective reenlistment bonuses as an across-the-board pay differential.

However, we note that most elements of compensation are common to the four military services. But the Army has been the most affected by the operations in Iraq and Afghanistan, resulting in negative shocks to recruiting and retention. The enlistment and reenlistment bonus programs provide the Army with an adjustment mechanism that obviated the need for compensation adjustments that were not service-specific. That a high percentage of enlistees and reenlistees received a bonus need not be viewed as an unnecessary use of bonuses if, in fact, their high use is in response to an overall, service-level shortfall of enlistments and reenlistments.

Furthermore, our comparisons indicate that even when the majority of Army enlistees and reenlistees received a bonus, there was substantial variation in bonus amounts and prevalence across occupations and lengths of enlistment (or reenlistment). This variation provides evidence that bonuses were used flexibly by the Army both to channel recruits into different occupations and service lengths and to expand the market and meet retention goals.

Specifically, in the case of enlistment, Army occupational specialties, such as infantry, field artillery, and air defense, consistently received substantially larger enlistment bonuses than other occupational areas. For example, in FY 2008, fire support specialists (13F) received an average bonus of $18,700 whereas armament repairers received a bonus of $2,800 on average. We also find variation across term of enlistment. Specifically, the premium for a six-year enlistment (relative to a five-year enlistment) was about $4,200, and for a five-year enlistment (relative to a four-year enlistment) it was about $2,300. In the case of reenlistment, the Army used complex rules to fine-tune the targeting of the dollar amount of SRBs at specific groups. The amounts of the SRBs depended on the occupation, rank, length of reenlistment of reenlistees, as well as on their skill (within an occupation), location, unit assignment, and deployment status.

Finally, the Army adjusted bonuses over time as conditions changed. As evidenced by the number of SRB program changes the Army announced each year in the FY 2001–2008 period, it seemed to manage its SRB program proactively. Furthermore, as shown by SRB reductions in the FY 2002–2003 time frame, and the substantial reductions it announced in March 2008, the Army is not reticent to reduce SRBs when retention is above target.

As with the Army, Marine Corps bonuses increased in terms of both the percentage of reenlistees receiving a bonus, from over 20 percent to about 80 percent for those at the first reenlistment point between FY 1996 and FY 2007, and the average SRB multiplier at the first term for those receiving a bonus, from over 2 to nearly 4. In the Navy, the percentage of reenlistees at the first term receiving an SRB varied over this period, increasing from around 60 percent in FY 1996 to 80 percent in FY 2000, but then declining to around 60 percent in FY 2007. On the other hand, average SRB multipliers among those receiving an SRB declined in the Navy between FY 2000 and FY 2007, from about 3.25 to 2. The Air Force experienced even larger swings. The percentage receiving an SRB at the first term increased from 20 percent to 80 percent between FY 1996 and FY 2002, falling back to less than 20 percent by FY 2007. Unlike the Navy, the average SRB multiplier in the Air Force for those receiving one increased from about 1.5 in FY 1996 to about 3.5 in FY 2007. Thus, in the Air Force, relatively few received a bonus in FY 2007 but those who received one had a substantially higher bonus than those in early years.

The large variations in bonuses over time in each service indicate that this compensation tool, unlike basic pay and the various allowances paid by the services, can be turned on and off relatively easily and quickly. Bonuses are used flexibly to respond to recruiting and retention changes. Furthermore, variations across occupations, and even across locations, skill subsets, units, and deployment status in the case of reenlistment bonuses, suggest that the services, notably the Army, used bonuses to target resources, even when the most personnel were receiving some sort of bonus.

Cost-Effectiveness (Efficiency)

We assess whether the bonus programs are too large or too small by comparing the cost of expanding these programs (and also the person-years of experienced personnel) using EBs and

SRBs with the cost of doing so using other compensation or personnel policy tools, notably pay. As discussed under the "Caveats" subsection, we do not assess whether the absolute levels of bonuses were optimal.

We estimate that enlistment bonuses are more cost-effective than pay but less cost-effective than recruiters as a way to expand the market for the Army. The marginal cost of enlistment bonuses (i.e., the cost per additional high-quality recruit) is estimated to be $44,900, compared to $57,600 for pay. We estimate a lower marginal cost for Army recruiters, $33,200. It is likely that we overstate the total marginal cost of bonuses. First, we account only for the market expansion effect and not the skill-channeling effects of bonuses. Second, our study finds that enlistment bonuses have a small but statistically significant effect on reducing attrition, thereby increasing the number of person-years provided by a given recruit, and this improvement in person-years is not included in our marginal cost estimate. Third, bonuses may induce enlistees to choose longer enlistment terms, producing more person-years. Again, we do not account for this effect in our estimate of cost-effectiveness. It is also important to note that although our cost-effectiveness estimate for recruiters is lower than it is for bonuses, the estimate does not account for any benefits associated with services' flexibility to target recruiters to specific regions of the country, or the costs associated with the time lag involved in expanding the recruiter force because of training time. These considerations regarding our cost-effectiveness measures indicate that although cost-effectiveness is one criterion for comparing recruiting resources, other considerations may also be important.

In the case of reenlistment, we provide a range of estimates of the marginal cost of SRBs using alternative assumptions and using different SRB estimates, depending on whether the SRB multiplier is conditional on deployment and whether we use estimates from the Army-only analysis or the four-service analysis. The estimates account for both the effects of SRB multipliers on reenlistment and the length of reenlistment. Our estimates for the Army indicate that the marginal cost of a change in the SRB multiplier at the first reenlistment point is in the range of $8,300 to $24,900 per person-year. For the second term, the estimate is in the range of $10,400 to $23,900 per person-year.

Our estimated marginal cost of first-term reenlistment bonuses for the Marine Corps is about the same as for the Army, a little higher for the Navy (in the range of $24,700 to $28,000), and substantially higher for the Air Force (in the range of $67,400 to $70,200). The marginal cost of reenlistment bonuses are usually higher for all services at the second term than at the first, varying from about $38,000 for the Navy to as high as $75,000 for the Marine Corps and $112,000 for the Air Force.

We note that our cost estimates of reenlistment bonuses incorporate the estimated effects of bonuses on length of reenlistment term. To the extent that the effects are positive but decreasing at higher multiplier levels and reflect the effects of bonus ceilings or the limited flexibility to choose term lengths, the cost-effectiveness of bonuses could be increased. The services could manage bonus ceilings more actively and, if necessary, increase the flexibility available to members to choose term lengths. Both of these steps would increase the cost-effectiveness of bonuses at the end of the second term.

For several reasons, bonuses are always likely to be more cost-effective than across-the-board increases in military pay: They can be targeted at occupations and zones, can be applied to a given interval of service (the reenlistment period), and can vary in amount. An across-the-board pay increase applies to all occupations, not just those with an impending shortage; creates a higher pay floor, which might mean higher pay costs in all future years; and gives the same pay increase to everyone. Military pay must be kept competitive overall, and pay increases provide the foundation for competitiveness. Bonuses allow for selective increases to differentiate pay by occupation and experience level and can be easily increased or decreased depending on current conditions. The estimated bonus costs at the end of the first term are likely to be substantially less than the marginal cost of raising military pay to achieve reenlistment goals, especially for the Army, Navy, and Marine Corps. Similarly, these costs are likely to be substantially less at the end of the second term for the Army and Navy relative to the marginal cost of raising pay. Since pay must be raised for everyone, not just reenlistees in critical occupations, the rents[3] associated with changes in pay are large and substantially more than the rents associated with reenlistment bonuses. That said, whether bonuses were optimally managed or set too high for too long is an open question.

We note that the marginal cost estimates for the Air Force are quite high, in contrast to the other services, as they are for the Marine Corps at the end of the second term. The high marginal costs come from the small bonus effects discussed above. A full understanding of why these bonus effects are small will require further research. Taken literally, they suggest that bonuses are a costly way to obtain additional person-years for these services at these reenlistment points. However, we urge caution in drawing this conclusion. First, our bonus effects may be biased downward and may be affected by bonus caps and limited flexibility in choosing LOR, as discussed above, and estimated effects that are too small lead to marginal cost estimates that are too high. Second, cost-effectiveness must be measured relative to an alternative approach to achieving reenlistments and, as noted in the previous paragraph, bonuses are more cost-effective than an across-the-board pay raise.

Other Results

Although these topics are not the focus of our analysis, we also report estimates of the effects of enlistment bonuses on Navy high-quality enlistments, the effects of EBs on Army first-term attrition, the effects of the Iraq War on Army high-quality enlistments, and the effects of deployment and stop-loss on the probability of Army reenlistment.

We find that enlistment bonuses have a smaller effect on Navy than on Army high-quality enlistments. Unlike the Army, the Navy did not expand its enlistment bonus program in recent years, and the percentage of Navy recruits receiving a bonus declined. On the other hand, average Navy enlistment bonuses differed substantially across occupational areas, suggesting that EB played more of a skill-channeling than a market expansion role for the Navy in recent years.

As noted above, we estimate that enlistment bonuses have a small but positive effect on the probability that an Army recruit completes his or her first enlistment term. A priori, we are not able to predict whether bonuses should increase or decrease attrition. On the one hand, larger bonuses are paid in installments, so recruits have a strong incentive to remain in ser-

---

[3]   Rent is a term from the economics literature. In the case of bonuses and enlistment and reenlistment bonuses, it refers to situations where bonus payments must be increased to expand enlistments or reenlistments over and above current enlistment and reenlistment levels. Consequently, individuals who would have enlisted or reenlisted in the absence of the increase earn a higher bonus than the one needed to induce them to enlist or reenlist. For example, suppose, hypothetically, that 100 high-quality recruits would have enlisted at a bonus of $5,000, but the Army needs to raise enlistment bonuses to $7,500 to increase enlistments to 108, then 100 high-quality recruits receive a rent of at least $2,500 because they received a bonus of $7,500 although 100 individuals would have enlisted for $5,000. Stated differently, these 100 individuals each receive a payment of at least $2,500 above their opportunity cost of joining.

vice to ensure collecting the full amount of their bonus. On the other hand, bonuses attract recruits who have less taste for military service and who, in the absence of bonuses, would not have enlisted. Consequently, we might expect an increase in bonuses to be associated with a lower probability of completing the first term. Our analysis indicates that, on net, the positive incentive effect of bonuses outweighs the negative effect of lower average taste for service on the probability of completing the first term for Army recruits.

We also find that operations in Iraq and Afghanistan have a negative effect on Army high-quality enlistments, but the size of the effect depends on our mathematical specification of the effects of these operations. When we represent these effects in terms of casualties, we estimate the effect of the Iraq War to reduce high-quality enlistments by an average of 6 percent. However, when we measure the effects of the Iraq War as the change in high-quality Army enlistments after the first quarter of FY 2003 that is unexplained by our model, we estimate a much larger effect. Using this method, we estimate that by FY 2006, the war accounted for a 50 to 60 percent decline in high-quality enlistments. The differences in these estimates (6 percent versus 50 to 60 percent) indicate the inherent difficulty of measuring the effects of national policy changes in a model estimated with aggregate data, as we use in this study. Nonetheless, the different approaches are consistent in indicating that the war had a sizable negative effect on high-quality Army enlistments, although the magnitude remains somewhat uncertain.

Some of the key results of our analysis of Army reenlistment relate to deployment and stop-loss. The Army imposed stop-loss on a significant fraction of its enlisted force, and we find that those subject to stop-loss were less likely to reenlist. However, we also find that the reenlistment rate at Zone A of those subject to stop-loss was only about two-thirds the rate of those who were not subject to stop-loss. In other words, only a third of the soldiers under the stop-loss policy would have exited from the Army if they had been permitted to do so. The remaining two-thirds of soldiers were willing to reenlist, even though they were under the stop-loss policy.

As with past studies, we find that deployed soldiers reenlist at a higher rate than nondeployed soldiers. On the other hand, more cumulative deployment time reduces the probability of reenlistment. Specifically, we find that a soldier with between one and two years of deployment is about 17 percentage points less likely to reenlist as a soldier with no deployment time.

Concluding Thoughts and Directions for Future Research

The main conclusion of this study is that enlistment and reenlistment bonus programs were important in helping the Army meet its recruiting and retention objectives and that the Army managed these programs flexibly by targeting bonuses to specific groups and adjusting them in a timely manner. At a time of war when some in Congress have called for the reinstatement of the military draft (Waller, 2003), bonuses were an important tool. They enabled the Army to respond to the adverse shocks to enlistment and reenlistment that resulted from declining civilian youth attitudes toward the military, frequent and long deployments, and rising end strengths. We also find evidence that the other services used bonuses flexibly by changing bonus amounts as retention shifted, and that reenlistment bonuses proved cost-effective relative to pay generally for the other services as well.

A remaining question is whether bonus levels were set correctly and, more broadly, whether the services could improve the management of these programs. Ideally, information is needed on the effects of enlistment and reenlistment bonuses on the supply of personnel to different occupations and an analysis of whether a different mix and different levels of bonuses than those actually observed would have resulted in more enlistments and reenlistments for the same cost. Such an analysis was beyond the scope of the current study but should be pursued in future research. Our analysis does provide some information on where management of bonuses might be improved. Specifically, we find that higher bonus multipliers reduce the length of reenlistment in some cases. That is, reenlistees choose shorter term lengths. We do not have the data to test alternative explanations for these results, but we suggest that the services might better manage bonus caps and provide members with the flexibility to choose longer terms to give them an incentive, and the ability, to choose longer reenlistment terms.

We recommend several areas for future research. In the case of enlistment, an empirical assessment is needed of how enlistment bonuses affect recruits' decisions to choose one occupation over another. With estimates of the skill-channeling effects of enlistment bonuses, one could then simulate how changing bonus differentials across occupations would affect enlistment rates into different occupations. In the case of both enlistment and reenlistment, we strongly recommend that Congress and DoD conduct an experimental test of the effectiveness of bonuses. Such a test would complement the results presented here by providing estimates without the confounding effects of reverse causality bias and would also permit estimates of the skill-channeling effects of bonuses on enlistment.

Acknowledgments

We are grateful to the U.S. Army Accessions Command for its help in answering questions and providing data. Specifically, we would like to thank Kevin Lyman, Don Bohn, and LTC Gregory Lamm. We are also grateful to John Noble at the Navy Recruiting Command for his support of our project. We received help and data from the Navy Recruiting Command, and we would like to specifically thank Jennifer Kelley and Michael Evans. We are indebted to personnel at the Defense Manpower Data Center for providing data, including Andrea Dettner, Jesica Kopang, Hannah Shin, Scott Seggerman, Ezekiel Budda, Christina Schmunk, Teri Cholar, and Peter Cerussi. Christopher Whaley provided excellent research assistance in preparing the Current Population Survey data we used. We also thank the many individuals at RAND who provided comments and observations in the RAND Defense Manpower Seminar and participants of the Western Economics Association meetings. We would like to also thank Stan Cochran in the Office of Officer and Enlisted Personnel Management for coordinating the acquisition of the data related to deployment and stop-loss. We are grateful to Richard Buddin and Paul Hogan for their reviews of an earlier draft of this report and to Margaret Harrell for guiding the RAND review process. We are also grateful to Robert Simmonds in the Office of Officer and Enlisted Personnel Management and to Steven Galing in the Office of Compensation for providing comments on our work. Finally, we are deeply grateful for the support of our project sponsor, Curtis Gilroy, Director, Accession Policy, Office of Accession Policy within the Office of the Under Secretary of Defense (Personnel and Readiness) and for the help we received from those who work in his office, especially Dennis Drogo, John Jessup, Christopher Arendt, and Robert Clark.

Abbreviations

AFQT	Armed Forces Qualification Test
AOS	additional obligated service
AVF	All-Volunteer Force
CONUS	continental United States
CPS	Current Population Survey
CSRB	Critical Skills Retention Bonus
CZTE	combat zone tax exclusion
DFAS	Defense Financial Accounting Service
DMDC	Defense Manpower Data Center
DoD	Department of Defense
DSRB	deployed selective reenlistment bonus
EB	enlistment bonus
ESRB	Enhanced SRB
ETS	expiration of term of service
FY	fiscal year
GAO	Government Accountability Office
GWOT	global war on terrorism
IMR	Inverse Mills Ratio
IV	instrumental variable
LOR	length of reenlistment
LSRB	location selective reenlistment bonus
MBP	monthly basic pay
MGIB	Montgomery GI Bill

MOS	military occupational specialty
OEF	Operation Enduring Freedom
OIF	Operation Iraqi Freedom
OSD	Office of the Secretary of Defense
PaYS	Partnership for Youth Success
RD	regression discontinuity
RMC	regular military compensation
S&I	special and incentive (pays)
SN	seaman
SQI	Special Qualification Identifier
SRB	selective reenlistment bonus
SRBM	Selective Reenlistment Bonus Multiplier
TSRB	targeted selective reenlistment bonus
YOS	year of service

CHAPTER ONE

Introduction

In fiscal year (FY) 2005, the Army failed to enlist enough individuals to meet its recruiting target for the year, and between FY 2005 and FY 2008, the percentage of Army recruits with at least a high school diploma fell below the target set by the Department of Defense (DoD). DoD requires that at least 90 percent of recruits be high school graduates and that at least 50 percent achieve a score of 50 or higher on the Armed Forces Qualification Test (AFQT). Although the quality of Army recruits fell below the first target, the Army was able to meet DoD's target AFQT scores. In contrast, the other services have consistently met both of these targets since FY 2000.

The Army and the Marine Corps have sought to maintain retention in the face of frequent and long deployments in support of operations in Iraq and Afghanistan. Both services have consistently met their overall annual enlisted retention goals since FY 2000.

To enable the services to meet recruiting and retention objectives during a time of war, the financial incentives associated with serving in the military increased substantially. Over the FY 2001–2006 period, military pay rose over 10 percent more than the earnings of comparably educated civilians (Simon and Warner, 2007). Special and incentive pays, especially those related to deployment, were also increased substantially. In FY 2008 dollars, the selective reenlistment bonus (SRB) budget for active duty personnel increased across DoD from $872 million in FY 2003 to $1.4 billion in FY 2008. The bulk of the increase was due to increases in the Army and Marine Corps budgets. In FY 2003, the Army and Marine Corps budgets were 21 percent of the DoD SRB budget. By FY 2008, these two services' SRB budgets were 66 percent of the DoD budget. The large ramp-up in enlistment bonuses began in FY 2006. In FY 2005, the DoD enlistment bonus (EB) budget increased from $296 million (in FY 2008 dollars) to $475 million in FY 2006, ultimately reaching $611 million in FY 2008. These increases enabled the services to reach their recruiting and retention goals during operations in Iraq and Afghanistan and at a time when the Army and Marine Corps were increasing endstrength.

In the face of these large increases in bonuses, Congress has questioned what the services have achieved with this money, whether bonuses have been effective in generating enlistments and reenlistments, and whether the services, especially the Army, have used these bonuses in a cost-effective manner.

In the case of enlistment bonuses, research from a 1980s enlistment bonus experiment found that bonuses have a modest market expansion effect (i.e., they increase high-quality enlistments relatively modestly) but they have a substantial skill-channeling effect (Polich, Dertouzos, and Press, 1986). That is, they are quite effective in inducing qualified youth who enlist to choose such critical skill areas as combat arms over less critical areas. Consequently, the traditional role of enlistment bonuses has not been to expand the market but to target enlistments into priority areas.

More broadly, the reports of various commissions and study groups concerned with military compensation, such as the 7th, 9th, and 10th Quadrennial Reviews of Military Compensation, have argued that enlistment and reenlistment bonus programs are essential parts of a cost-effective military compensation system.[1] The military recruits and trains personnel to do a wide variety of tasks. Some tasks require little training but others require complex and expensive training. Personnel who have received complex and expensive training are more marketable in the civilian labor market than personnel who have not received such training. Furthermore, the characteristics of military jobs—including the conditions of work and exposure to the risk of injury or death—differ widely by military occupation, location, and service. Given the wide variation in military job types and job characteristics (what economists call "job heterogeneity"), it would be very expensive for the military to balance its detailed job requirements (i.e., demand) with the number of personnel willing to fill these requirements (i.e., supply) using a common pay table for all personnel. If all personnel were paid the same, military pay would have to be set at a high level to meet its requirements in the hard-to-fill positions. Such a system would be very expensive and would result in an overall excess supply of personnel.[2]

Over time, therefore, the military has developed numerous special and incentive (S&I) pays to manage the force. S&I pays, of which enlistment bonuses and selective reenlistment bonuses are two key parts, permit military force managers to balance demand and supply in detailed categories without the need to expand the elements of compensation that are common to all personnel.[3]

Recently, the Army has been using enlistment bonuses as a key tool to expand the market. As will be discussed in Chapter Two, about 70 percent of Army enlistees received an enlistment bonus in FY 2008. Thus, nearly three out of four recruits in FY 2008 received a bonus. If bonuses were used only for skill-channeling, one would not expect a large segment of enlistees to receive a bonus, only those who enter critical skill areas. The fact that bonuses were nearly universal has raised questions by Congress and the Government Accountability Office (GAO) about the cost-effectiveness of the Army's enlistment bonus policies.

The Army offered reenlistment bonuses to most members who reenlisted in FY 2006. Because reenlistment bonuses were nearly universal, GAO and Congress have also raised questions about whether such bonuses were paid to personnel who would have reenlisted in the absence of these bonuses, thereby reducing their cost-effectiveness.

Congress has directed DoD to provide a report that describes bonuses—specifically, the number used for recruiting and retention, the average amount provided for each military occupational specialty (MOS), the length of contract required for each type of enlistment and reenlistment bonus, and metrics of performance to determine their effectiveness. To address questions about the effectiveness of bonuses and their cost relative to other resources, models

---

[1] See, for example, Chapter III of the *Report of the 9th Quadrennial Review of Military Compensation* (DoD, 2002).

[2] Personnel who are paid more than necessary to get them to do something are said to be earning "economic rents," as defined in a previous footnote.

[3] There has been long-standing debate in the Department of Defense and elsewhere about the extent to which the military should use S&I pays to manage the force. Some observers believe that all military personnel should be paid the same but other observers stress the high cost of such a philosophy. Hogan, Simon, and Warner (2004) discuss these philosophies in some detail.

of enlistment and reenlistment supply are required, estimated with data on enlistments and reenlistments, bonuses, and other factors that could affect enlistment and reenlistment.

Past studies have estimated the effects of bonuses on enlistment and reenlistment, but these studies do not use recent, post–FY 2005 data. Because of long and frequent deployments and the unusual stresses on the armed services since FY 2005 as a result of combat operations in Iraq and Afghanistan, it is possible that the effects of bonuses have changed in recent years. Furthermore, the change in U.S. strategy and the "surge" in forces in FY 2006 may have altered service members' views of military service, also implying changed responsiveness to bonuses. On the other hand, estimates of the effect of bonuses on enlistment and reenlistment from different periods of time since the beginning of the All-Volunteer Force (AVF) in FY 1973 seem not to be highly sensitive to the time period used to study bonus effects.[4] Thus, it is possible that estimated effects have not changed in recent years.

The research summarized in this report provides information to help DoD respond to the Congressional directive. The project estimated enlistment and reenlistment models, yielding estimates of the effects of bonuses and other factors on the enlistment supply of high-quality youth and on the reenlistment of qualified personnel at the first and second (also known as Zones A and B) reenlistment decision points.

In addition, this research estimates the effects of enlistment bonuses on the decision of service members to leave the service before the end of their first enlistment term. Bonuses might affect attrition for two reasons. First, enlistment bonuses are often paid in installments, so members have an incentive to stay in service to receive future installment payments. For this reason, enlistment bonuses may have a negative effect on attrition. On the other hand, bonuses might induce individuals to enlist who have a lower taste for the military or who have better civilian opportunities in the absence of bonuses. These individuals might be less likely to complete their first term, and so bonuses may be associated with higher attrition rates. Because these reasons yield opposite predictions about the effects of bonuses on attrition, we cannot predict a priori how enlistment bonuses will affect attrition. The empirical analysis we present in this report provides estimates of the net effect of these opposing forces.

In short, the research presented in this report addresses the following questions:

- What is the effect of enlistment bonuses on the supply of high-quality recruits?
- To what extent did increases in enlistment bonuses between FY 2004 and FY 2008 offset declines in Army high-quality enlistments over that period, and could more cost-effective recruiting resources have been used instead?
- What is the effect of enlistment bonuses on attrition? Do bonuses increase or reduce person-years and enlistments?
- What is the effect of reenlistment bonuses on reenlistments and the length of reenlistment?
- Are bonuses cost-effective relative to other resources?

In addition to addressing these questions related to bonus effectiveness, the project also provides information on the number of bonuses used for recruiting and retention, their average amount by military occupational specialty, and contract length, as required by the Congressional directive.

---

[4]   See Warner and Asch (1995) and Asch, Hosek, and Warner (2007) for a review of past bonus estimates.

To address the questions above, our project estimated enlistment, attrition, and reenlistment regression models. We estimate a model of Army high-quality enlistment supply using data from FY 1999 to FY 2008 provided by the Army. Individual enlistment data are aggregated by state and quarter, and we estimate an aggregate model of enlistment supply. In other words, we estimate a model of the (logarithm) of the number of high-quality Army enlistments per population as a function of variables that vary by state and quarter, including enlistment bonuses, recruiters, civilian pay and unemployment rate, and other factors.

We also estimate an aggregate regression model of Navy high-quality enlistment supply and provide estimates of the effects of Navy bonuses, recruiters, and other factors. We chose the Navy both because of the availability of data provided by the Navy and because the pattern of Navy bonuses differs from that of the Army (as will be discussed in Chapter Two). Thus, the Navy provides a contrasting example to the Army.

We estimate a model of Army reenlistment using data covering members making reenlistment decisions between FY 2002 and FY 2006 in 24 military occupational specialties constituting about half of Army entrants; these date were provided by the Army and by the Defense Manpower Data Center (DMDC). The reenlistment model is a model that formulates the individual member's decision to reenlist or leave the service, as a function of reenlistment bonuses, deployment experiences, and other factors. Thus, the reenlistment model is estimated at the individual level.

We estimate the reenlistment model at the individual level but the enlistment model at a more aggregated state level because, in the case of enlistment, we do not have data on individuals who choose not to enlist. In contrast, in the case of reenlistment, we have data on those who choose to reenlist and on those who choose to leave. Consequently, we take these different approaches.

We also estimate the effect of bonuses on reenlistment and on length of reenlistment for all four services, building on a recently published RAND report on reenlistments and deployments (Hosek and Martorell, 2009). This study estimated the effects of reenlistment bonuses and of deployment on reenlistments, using DMDC data for each service. We extend this study by refining the measure of selective reenlistment bonuses used, estimating the effects of bonuses on length of reenlistment, and providing marginal cost estimates. The data used for this part of the study also come from DMDC but were built independently.

The analysis that builds on the Hosek and Martorell report is presented separately from the results for the subset of 24 Army occupations because the Army-only analysis contains more detail about the Army's bonus-setting policies than the four-service analysis. In addition, the statistical methodology is somewhat different. Thus, comparing the estimates from the two separate analyses provides information about how sensitive the estimates are to the statistical tools used.

It is important to note that although we adhere to the convention of referring to our regression estimates as the "effects" of bonuses, our results measure associations between bonuses and enlistment, attrition, and reenlistment and do not represent causal effects. Unobserved factors correlated with both bonuses and these outcomes might not be captured in our models, such as local attitudes toward higher education, in the case of enlistment, and incentives to choose specific occupations, such as choice of location, in the case of reenlistment. Consequently, the relationship between bonuses and outcomes that we estimate may be due to a third factor, unobserved in our data, that drives both bonuses and the outcomes of interest. Furthermore, our estimates are not based on structural models of the decision to join the military or the decision to stay in the military whereby we estimate the underlying parameters that determine those decisions. Instead, we take a parsimonious, reduced-form approach where we isolate the effect of bonuses from the effects of other observed variables on enlistment, attrition, and reenlistment. The advantage of this approach is that it is simpler to implement and permits us to estimate the effects of many other variables on outcomes, such as the effects of other recruiting resources, in the case of enlistment, and the effects of deployment, in the case of reenlistments.

The report is organized as follows. Chapter Two provides background information on the trends in high-quality enlistments and enlistment bonuses for the Army and Navy. Chapter Three presents the econometric modeling approach we use and discusses the data we used in our enlistment analysis in more detail. Chapter Four presents our enlistment supply results for the Army and Navy, and Chapter Five presents the attrition results for the Army. Chapter Six gives background information on Army SRB policy in recent years, and Chapter Seven presents the Army reenlistment analysis. Chapter Eight extends the Hosek and Martorell (2009) reenlistment analysis and gives results for all four services. Chapter Nine offers our conclusions.

The report provides a compendium of information as well as reference material from which policymakers in DoD can draw to address the questions raised by Congress. As such, the report is quite comprehensive. Some readers may be interested only in specific topics. The chapters are generally self-contained, and some readers may choose to read selected chapters rather than the entire document.

CHAPTER TWO

Background on Enlistment Bonuses

This chapter reports recent trends in enlistment bonuses. Congress directed DoD to provide information on the number of bonuses, the average amount by military occupational specialty, and their differences by contract length. This chapter summarizes this information with respect to enlistment bonuses. (More detailed information is provided in Appendix A and Chapter Five provides similar information for reenlistment bonuses.)

Figure 2.1 plots the share of all Army enlistees who received cash enlistment bonuses, by month, from FY 2000 to FY 2008. After increasing during FY 2000 and FY 2001, the share of enlistees receiving bonuses remained stable for the next three years at around 40 percent. Since FY 2005, the share of enlistees with cash enlistment bonuses has grown, fluctuating most recently between 60 and 80 percent.

Figure 2.2 plots the monthly average amount of the contracted bonus based on contract date for those offered a bonus. Bonus amounts are expressed in FY 2008 dollars. Army enlistment bonuses actually fell in value in real terms between FY 2001 and FY 2004, as the Army generally scaled back investments in recruiting resources. Perhaps the most remarkable trend, however,

Figure 2.1
Percentage of Army Gross Contracts That Received an Enlistment Bonus

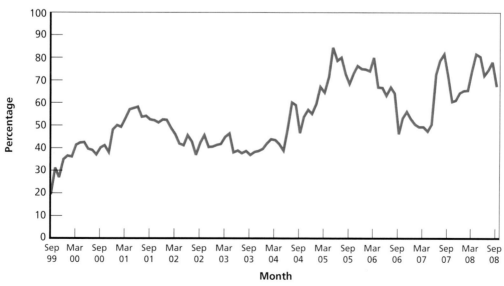

SOURCE: Authors' calculations based on Army recruiting data.
NOTE: Gross refers to total contracts, not net of any attrition or dropouts.
RAND *MG950-2.1*

Figure 2.2
Average Monthly Army Enlistment Bonus, Conditional on Receiving a Bonus (in FY 2008 dollars)

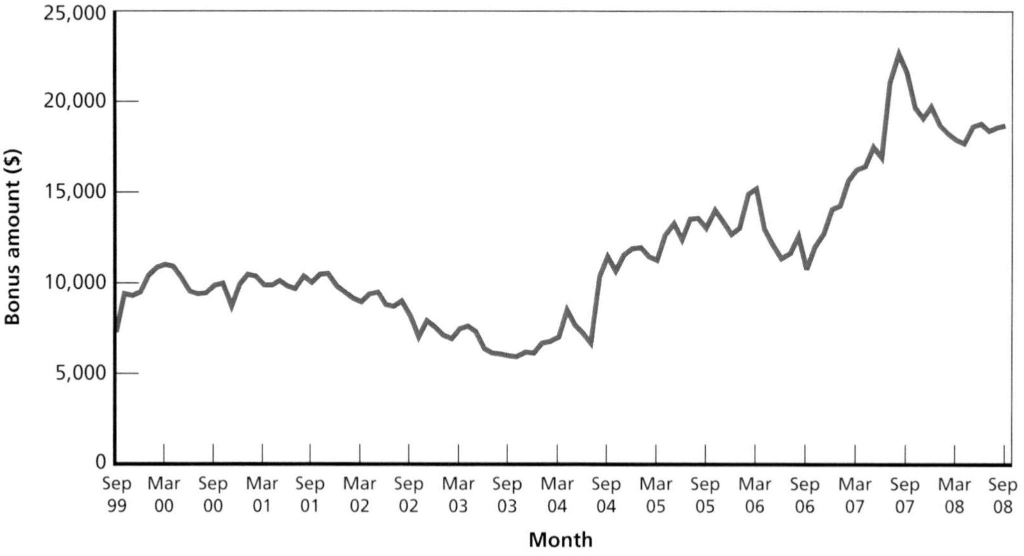

SOURCE: Authors' calculations based on Army recruiting data.
RAND *MG950-2.2*

is the dramatic increase in enlistment bonuses starting at the end FY 2004 and beginning of FY 2005. The increase occurred in two major phases—an almost doubling of average bonuses from around $5,600 at the start of FY 2004 to $10,500 at the start of FY 2005 and then a doubling of bonuses during FY 2007. When the Army offered the highly publicized "quick-ship" bonuses in summer 2007, as a way to ensure that it would meet its FY 2007 contract mission, average bonuses peaked at about $22,400. By mid-2008, average bonuses were about $18,000, roughly three times the levels of FY 2003 and FY 2004. In summary, the Army substantially expanded both the availability and magnitude of bonuses between FY 2005 and FY 2008.

As demonstrated by Figure 2.3, the types of bonuses employed by the Army have changed somewhat over time. Figure 2.3 depicts the percentage of cash enlistment bonuses paid by the Army that fall into several broad categories.[1] Seasonal bonuses include the aforementioned quick-ship bonuses and are linked to the timing of entry into basic training. The quick-ship bonus is given to high-quality recruits who enlist in selected occupations and agree to ship to boot camp within a relatively short time frame—between one to two months. PaYS bonuses are bonuses paid through the Army's Partnership for Youth Success program, which typically combines a cash enlistment bonus with a guarantee from a civilian employer to offer preferential hiring consideration to enlistees with particular occupational skills on completion of their first term of service. HiGrad bonuses are bonuses tied to educational credentials, including college credits. Cash bonuses are general bonuses typically linked to critical occupations or enlistment contract length. About one-fifth of bonuses fall into other, smaller programs, such as cash bonuses paid to those with civilian skills that the Army requires and bonuses paid to those with prior military service. Also included in the "other" category are bonuses that were paid in conjunction with the Army College Fund, the Loan Repayment Program, and the Army Advantage Fund.

---

[1]   The share calculation is based on the number of contracts; dollar-weighting the share yields comparable insights.

Figure 2.3
Percentage Distribution of Army Contracts Across Bonus Types

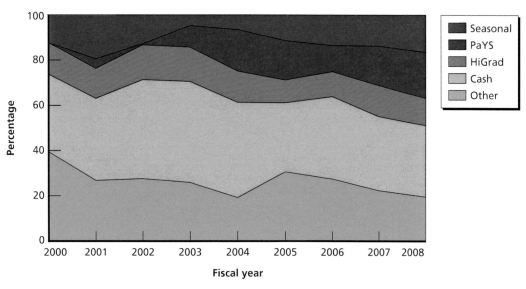

SOURCE: Authors' calculations based on Army recruiting data.
RAND *MG950-2.3*

Since FY 2003, the seasonal and PaYS-related bonuses expanded substantially, with the combined share of these bonuses growing from 14 percent in FY 2003 to over 35 percent by FY 2008. The share of enlistees receiving bonuses for educational credentials has remained relatively constant since FY 2000, and the share of bonuses associated with niche programs has fallen somewhat.

Conceptual discussions of enlistment bonuses typically differentiate two distinct purposes of bonuses—market expansion and skill-channeling. Enlistment bonuses expand the market by inducing individuals who otherwise would be disinclined to join the military to enlist to earn a bonus. Bonuses can also induce individuals who were already planning to enlist to enlist into occupations that are of particular value to the services. Bonuses are thus used to both increase the size of the entering force and ensure that the skill composition of the entering force meets the military's requirements. As discussed in Chapter One, past research from the enlistment bonus experiment provides evidence that bonuses have traditionally been highly effective in their skill-channeling role but less effective than recruiters and other resources in the market expansion role. (We provide new evidence on the size of the market expansion effects of bonuses in Chapter Four.)

Examining patterns of bonus receipt within occupations permits us to provide evidence on the market expansion versus skill-channeling roles of enlistment bonuses in the Army over time. Figure 2.4 plots the average bonuses in FY 2008 dollars for a representative set of occupational fields within the Army, by fiscal year. (Averages across all occupational fields are reported in Appendix A.) Several interesting patterns are apparent. Occupations such as field artillery and infantry, which traditionally have received bonuses and which obtained bonuses of several thousand dollars even before 9/11, saw substantial increases in bonuses, with averages growing beyond $20,000 for some occupations. However, also notable is that fields that have historically received few bonuses, such as law enforcement and religious services, began receiving bonuses beginning in FY 2005. The gaps across occupations through the sample

Figure 2.4
Average Army Enlistment Bonus, by Selected Occupational Areas (in FY 2008 dollars)

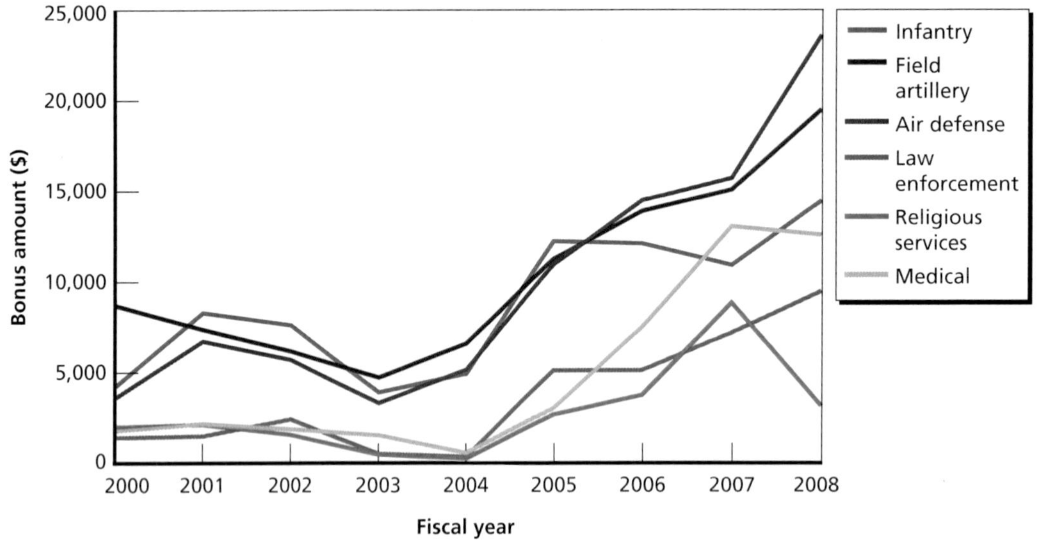

SOURCE: Authors' calculations based on Army recruiting data.
RAND *MG950-2.4*

period indicate an important skill-channeling role for bonuses, but the fact that large increases in bonuses since FY 2005 have occurred across almost all occupations indicates that bonuses have been increasingly used as a market expansion tool.

To what extent were bonuses targeted at those who contracted for longer enlistment terms? Figure 2.5 reports the bonus premium associated with an enlistment term of four, five, or six years relative to three years, the modal enlistment term. The reported premia are calculated from a regression of average bonuses on term length, controlling for the occupational composition of recruits. This approach allows us to account for the fact that some occupations feature both longer enlistment terms and different bonus levels. For example, the plotted value of $3,526 for the four-year enlistment in FY 2000 indicates that, on average, a person who enlisted for four years in FY 2000 received a bonus that was $3,526 higher than someone enlisting in the same occupation for only three years. Similarly, in FY 2000, those enlisting for five years received $4,751 more than those enlisting for three years, and those enlisting for six years received $7,027 more than three-year enlistees. Although the premium for moving from a four- to a five-year enlistment has remained relatively stable throughout the data period, the six-year premium grew between FY 2004 and FY 2008. The Army offered about $1,500 more for the longest enlistment term between FY 2006 and FY 2008 than it did between FY 2000 and FY 2003.

To summarize, we see evidence of a substantial increase in enlistment bonuses in the Army between the end of FY 2004 and FY 2008, reflecting both an expansion of bonus eligibility and an increase in the average size of bonuses. Bonus growth has occurred across almost all occupations, suggesting that bonuses have increasingly become a market expansion tool in the Army. Nevertheless, occupational differentials still exist, suggesting that bonuses continue to be used as a skill-channeling tool. The Army now offers a larger premium for six-year enlistments than it has historically. Within bonus types, seasonal bonuses and bonuses tied to the Partnership for Youth Success program have increased in prevalence.

Figure 2.5
Increase in Bonuses Relative to a Three-Year Enlistment Bonus, by Term of Service, Army
(in FY 2008 dollars)

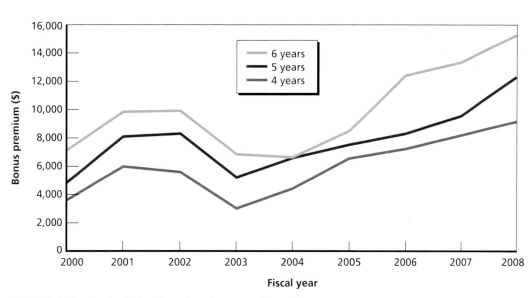

SOURCE: Authors' calculations based on Army recruiting data.
NOTE: Bonus premia have been regression-adjusted to control for differences in the occupational mix of
enlistees with different terms of service.
RAND *MG950-2.5*

Given the changes in Army enlistment bonuses demonstrated in Figures 2.1 through 2.5, a natural question arises regarding the extent to which similar changes were implemented in other services. Figure 2.6 plots the percentage of new contracts that included a cash enlistment bonus in the Navy. Although fluctuating somewhat before FY 2004, the percentage of new Navy enlistees receiving bonuses has declined more recently, from over 70 percent at the beginning of FY 2004 to approximately 25 percent by the end of the sample period. This pattern contrasts with that for the Army, which increased eligibility for bonuses substantially starting in FY 2004.

The contrast is less apparent when examining average bonus amounts among those receiving bonuses in the Navy (Figure 2.7). After several years at the $6,000 level, average Navy enlistment bonuses rose steadily beginning in FY 2005, peaking above $14,000 by the end of the sample period. Thus, in both services, average bonus amounts have grown appreciably in recent years, but in the Navy, higher bonus amounts have been associated with reduced availability, whereas in the Army both amounts and availability have expanded.

Figure 2.8 depicts the trend in enlistment bonuses for a representative set of Navy occupations. (Average bonuses for the full set of occupations are reported in Appendix A.) For the Army, enlistment bonuses grew across the full spectrum of occupations beginning in FY 2004, suggesting that market expansion was a key role for bonuses. For the Navy, there continues to be considerable idiosyncratic variation across occupations in the level and trajectory of bonuses. For example, the generic "seaman" rating (SN), one of the most common enlistment classifications, saw a large, temporary increase in average bonuses in FY 2006. Bonuses for cryptologic technicians grew steadily between FY 1999 and FY 2008, and bonuses for some ratings, such as hospital corpsman, have declined in recent years. The absence of a systematic pattern in Navy bonuses suggests that bonuses continue to be used primarily for

Figure 2.6
Percentage of Navy Gross Contracts That Received an Enlistment Bonus

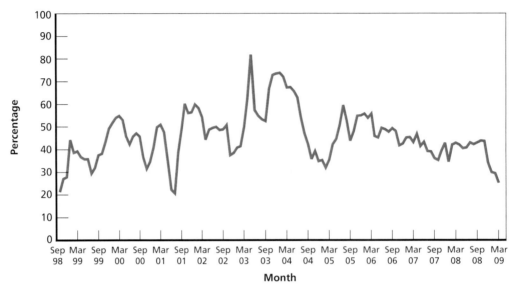

SOURCE: Authors' calculations based on Navy recruiting data.
NOTE: Gross refers to total contracts, not net of any attrition or dropouts.
RAND *MG950-2.6*

Figure 2.7
Average Monthly Navy Enlistment Bonus, Conditional on Receiving a Bonus (in FY 2008 dollars)

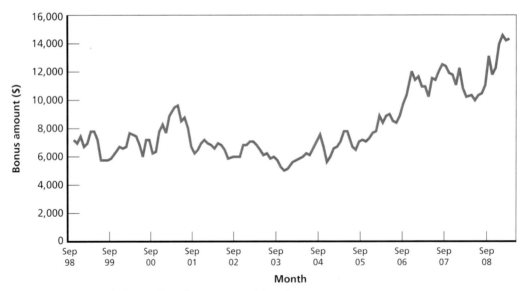

SOURCE: Authors' calculations based on Navy recruiting data.
RAND *MG950-2.7*

Figure 2.8
Average Navy Enlistment Bonus, by Selected Occupations (in FY 2008 dollars)

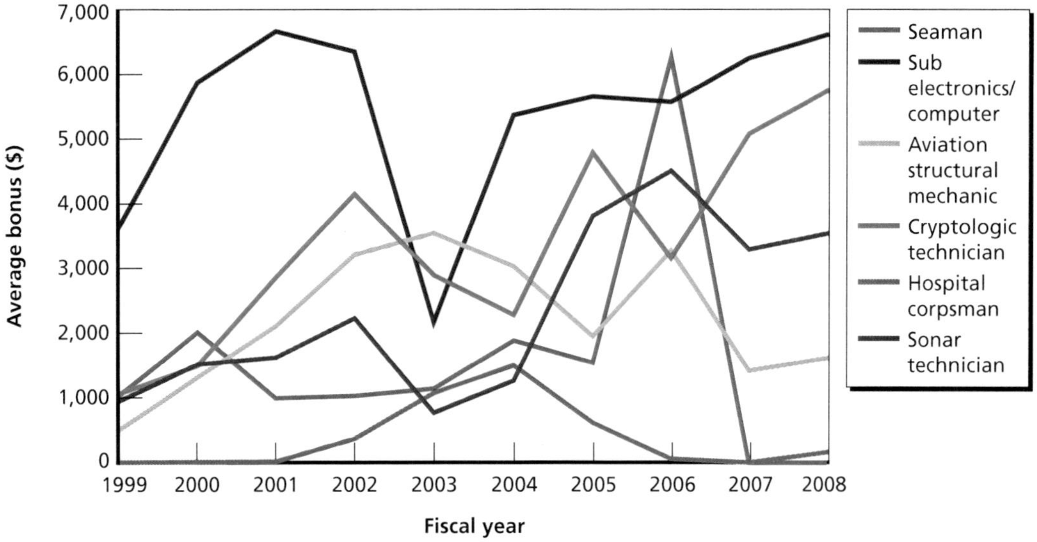

SOURCE: Authors' calculations based on Navy recruiting data.
RAND *MG950-2.8*

skill-channeling in this service, allowing the Navy to shift recruits across occupations to meet its workforce composition needs.

Figure 2.9 plots the bonus premium associated with an increase in enlistment term from four to five or four to six years, controlling for occupation. These premia have been calculated using the same method as those in Figure 2.5. After rising between FY 1999 and FY 2001, bonus premia for longer enlistment terms remained remarkably stable between FY 2002 and FY 2005 before rising beginning in FY 2006 for six-year terms and in FY 2007 for five-year terms. During the bulk of the sample period, five-year enlistees received roughly $4,000 more than four-year enlistees, and six-year enlistees received $6,000 more than four-year enlistees.

In summary, for the Navy we see evidence of an expansion of bonus amounts, but only for selected occupations. Although the Army has increased bonuses generally since FY 2004, bonus increases in the Navy have been more targeted. In recent years, Army bonuses have played both a market expansion and skill-channeling role and the Navy has continued to use bonuses for skill-channeling.

Figure 2.9
Increase in Bonuses Relative to a Four-Year Enlistment Bonus, by Term of Service, Navy (in FY 2008 dollars)

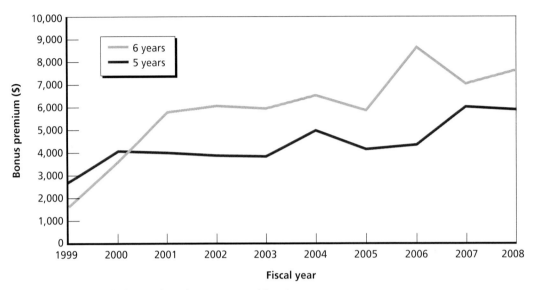

SOURCE: Authors' calculations based on Navy recruiting data.
NOTE: Bonus premia have been regression-adjusted to control for differences in the occupational mix of enlistees with different terms of service.
RAND *MG950-2.9*

CHAPTER THREE

Methodology and Data for the Enlistment Model

To understand how the large increases in enlistment bonuses documented in the previous chapter affected enlistments between FY 2005 and FY 2008, we require a model linking bonus amounts to the number of enlistments. In this chapter, we describe our statistical modeling approach. Our basic model relates fluctuations in the number of high-quality enlistments at the state/quarter level to changes over time in the resource, economic, and demographic environment within each state. Our earlier report (Asch, Heaton, and Savych, 2009) reviews the past literature on enlistment supply. We first describe our methodology and then the data used to estimate our models.

Methodology

We estimate the relationship between high-quality enlistments and enlistment bonuses using panel regression methods. We focus on high-quality enlistments because these are supply-constrained; ideally, the services would enlist only high-quality recruits, but they are forced to enlist some individuals who fall outside this classification.

Underlying our model is the economic theory of occupational choice built on the random utility model (McFadden, 1983). Individuals are assumed to choose to enlist if the military provides greater utility or satisfaction than the best civilian alternative. Factors affecting utility include the taste for military service versus civilian opportunities, the relative financial returns to military and civilian opportunities, as well as such random factors as health or economic shocks. We do not estimate a structural model of enlistment. Doing so would provide estimates of the underlying taste distribution for military service and the distribution of the random shock. Such models are usually estimated with data on the employment choices—military or civilian—of individual youth, usually with data from the National Longitudinal Surveys. However, such data are not available for recent cohorts of youth. Instead, we estimate what is known as a reduced-form model that posits that high-quality enlistments are associated with a set of variables that capture taste for military service, such as demographic factors, the financial return to military service and civilian opportunities (such as enlistment bonuses and pay) and factors that can affect the taste for military service, such as recruiters. These models and the past studies that have estimated them are reviewed in Asch, Heaton, and Savych (2009).

Because we estimate a reduced-form model that measures associations between the variables in the model, it does not represent causal effects. Clearly, there may be unobserved factors correlated with both bonuses and enlistments that are not captured in our model, such as local attitudes toward higher education. A proper causal analysis would require that we isolate the effects of bonuses from all confounding factors, which is infeasible in the present setting. At the same time, absent exogenous variation in resources, the relationships uncovered by the model provide our best means for projecting the effects of changes in bonuses. Although in the discussion that follows we adhere to the convention of referring to our estimates as "effects" estimates, these caveats regarding causality should be borne in mind.

We perform our analysis at the state/quarter level. Our model is given by Equation (3.1).

$$Log\left(\frac{HQ\ \text{Contracts}}{\text{Population}}\right)_{it} = \beta X_{it} + \gamma_i + \theta_t + \varepsilon_{it} \tag{3.1}$$

We partition our covariates into those that vary both over time and across states, X_{it}, a vector of state fixed effects, γ_i, and a vector of time effects, θ_t. The final term is ε_{it}, the random effect that captures the omitted variables. It is not possible to identify variables that vary only over time or only across state. Although we have no variables that vary only across state, we have several variables of interest that vary only over time. These include the effects of operations in Iraq (as measured by casualties), the maximum Montgomery GI Bill (MGIB) benefit, and the logarithm of the ratio of regular military compensation to annualized civilian weekly wages, as we describe in more detail below when we discuss our variables.[1] Therefore, we estimated two specifications for the Army and for the Navy. The first specification includes state effects but not time effects. The second specification includes both state and time effects. In the first specification, we instead include a polynomial in time to control for general time trends. Variables that vary only by time are not identified in the second specification, and they are excluded in this specification but not the first.

In the first specification, the effects of such variables as pay and the MGIB program, which vary over time but not across states, are identified using changes over time in contracts at the national level. Given the relatively few number of time periods (36) in our sample and the infrequency with which military pay and the MBIG are altered, we caution that the effects of pay and the MGIB program are not likely to be well identified in this specification. Cleanly estimating the effects of national-level variables remains a general problem for aggregate enlistment supply models.[2] For explanatory variables that vary across both states and time, such as the unemployment rate, effects are identified by comparing high-quality enlistments across states with large changes over time relative to those with small changes and comparing aggregate changes over time. In the second specification with both time and state effects, the effects of such explanatory variables as bonuses, which vary both by time and location, are identified by comparing states that experience large changes over time in these variables to states that do not. This approach discards national-level variation, which precludes the inclusion of variables

---

[1] As discussed below, although there is some cross-sectional variation in the military/civilian pay ratio because of differences in average civilian pay across states, this variation is small relative to the variation over time generated by numerous legislative pay raises.

[2] An alternative approach to estimating coefficients for variables that vary over time but not across states is to use the estimate version of the model with a full set of time and state fixed effects and then regress the estimated time fixed effects on the time-varying covariates in a separate stage, as in Warner, Simon, and Payne (2001) and Asch, Heaton, and Savych (2009). We found that this approach yielded similar effect estimates to our first specification.

that do not vary across states (such as military pay) but renders the estimates less subject to bias from factors omitted from the model.[3]

Our approach has two key advantages. First, the state fixed effects mitigate bias that may arise as a result of the omission from the model of relevant factors at the state level that are relatively constant over time. For example, the model does not include variables that indicate the quality of the state's university system, which could negatively affect enlistments in a state. However, failure to account for this factor is unlikely to bias our estimates once we include state fixed effects. Second, by using two estimation approaches, we maintain the flexibility to estimate effects for the widest possible set of factors, recognizing the potential for omitted variable bias, or to focus attention on a narrower set of factors but employ an approach that is most likely to generate accurate estimates of effects.

Given that services compete for the same pool of high-quality youth, it seems possible that factors chosen by one service might affect enlistments in another. For example, increased numbers of Army recruiters in a particular location might draw recruits into the Army who otherwise would have joined the Navy, in which case Army recruiters would negatively affect Navy enlistments. Although, ideally, we might wish to allow for the possibility of such cross-service effects in our model, the small sample sizes afforded by an aggregate model (relative to a model incorporating individual-level data) coupled with the high correlation across services in some incentives, such as bonuses, hamper our ability to precisely estimate such interactions. We instead take the more straightforward route of separately estimating Equation (3.1) for each service. Thus, one limitation of our modeling approach is its inability to provide direct evidence regarding cross-service effects.

Variables and Data

To measure the dependent variables—high-quality enlistments in each quarter and state and by service—we use data on high-quality gross contracts from FY 2000 through FY 2008 provided by the U.S. Army and the U.S. Navy. The data were provided at the level of the individual enlistment, and we aggregated the data by state and quarter.

Our primary explanatory variables are enlistment bonuses and other measures of military recruiting resources, namely, recruiters, college funds, and pay.[4] We measure recruiters relative to the size of the adult population.[5] We measure enlistment bonuses using the average total cash enlistment bonuses offered to new recruits in a given state and quarter. We calculate aver-

---

[3] By bias, we mean that the parameter estimates in our model of different factors are not equal to the true effects of these factors.

[4] Some studies (e.g., Dertouzos and Garber, 2003) consider advertising expenditures as an additional explanatory variable of interest. Some recent research (Dertouzos, 2009) also suggests that groups may vary in their responsiveness to certain types of advertising, such as cable TV commercials. Unfortunately, service-level advertising data of sufficient quality were not available for the entire time period covered by the study, precluding the inclusion of advertising in our analysis. If advertising occurs in areas with positive growth of other resources, our estimates of the effects of these resources may be biased upward.

[5] Navy recruiter data were available only by recruiting district, and some recruiting districts cover areas in multiple states. We developed state-level measures of recruiters by allocating them to states on the basis of the population proportions of the areas covered by each recruiting district.

age bonuses from contract microdata provided to us by each service.[6] We measure service-level college fund availability as the proportion of new recruits in a given state and quarter who were offered the college fund.[7]

Because MGIB benefit generosity is determined at the national level, our MGIB measure enters the analysis in the second specification. We measure MGIB benefit generosity using the current maximum annual benefit level at the contract date, and we account for changes in the cost of schooling by denominating this measure using average college tuition.[8] Although measuring the MGIB using contemporaneous benefit levels is not ideal, given that benefits are not actually received until several years later, it seems reasonable to expect that increases in current benefits would affect perceptions of future benefits.[9]

Past researchers have demonstrated that military compensation is highly correlated with enlistments, as discussed in Asch, Heaton, and Savych (2009). Our pay measure is the ratio of regular military compensation to average civilian pay, where civilian pay has been calculated separately using the Current Population Survey (CPS). Regular military compensation includes basic pay, basic allowance for subsistence, basic allowance for housing, and the advantage associated with receiving these allowances tax-free. By constructing our measure as a ratio, we attempt to capture the financial attractiveness of military service relative to other types of employment. Basic pay is set military-wide and does not vary by duty station. Although there is some cross-sectional variation in the military/civilian pay ratio because of differences in average civilian pay across states, this cross-sectional variation is small relative to the variation over time generated by numerous legislative pay raises. We thus enter our pay variable in the first but not the second specification of our regression model and identify the effects of pay on recruiting using time-series variation.

We also incorporate a variable related to the presence of positive influencers, namely, the percentage of population in a quarter, over age 35, who are military veterans in the state. We expect that as the veterans' population declines in a state, high-quality enlistments should decline as well. Other than civilian pay, which is included in the ratio of military-to-civilian pay, we include factors related to external civilian opportunities, namely, the state's unemployment rate in a given quarter and the percentage of the state's high school graduate population between the ages of 17 and 30 enrolled in college. As these external opportunities improve (or decline), military service becomes less (or more) attractive, and we expect that high-quality enlistments should increase.

An additional variable included in the analysis relates to eligibility for enlistment, namely, the percentage of noncitizens. The United States has experienced appreciable increases in its noncitizen population over the past decade, particularly among Hispanic youth. To capture the

---

[6]   Because of a modification of the Army's information technology system in FY 1999, bonus data are unavailable for the Army during Q2 and Q3 of that year. For these quarters, we impute the average bonus amount over the rest of our sample.

[7]   The actual offer amounts would provide more information than our measure, but college fund amounts were not consistently coded in the microdata.

[8]   We obtained tuition data from the College Board's *Trends in College Pricing* publication series.

[9]   Although this measure is appealing for its simplicity, it does not account for the fact that the value of this incentive is ultimately related to expectations for utilization, which may vary for different individuals and are affected by rates of time preference and perceptions of the likely future benefit generosity. An alternative approach for measuring MGIB benefits would be to attempt to predict the likely use of benefits from observable individual characteristics and to adjust for the fact that benefits cannot be claimed until the future. In earlier stages of our analysis, we examined this possibility but found that it did not generate substantially different results from the simpler alternative.

potential effects of such changes on enlistments, we include the log of the noncitizen proportion of the population by state and year as an additional explanatory variable in our model. The theoretical relationship between citizenship and propensity to enlist is ambiguous—although noncitizens may be attracted by the expedited naturalization process provided to service members and may have higher levels of patriotism than the general population, the military may present larger language or other assimilation barriers than other types of employment.

We also attempt to quantify the effects of operations in support of the wars in Iraq and Afghanistan on enlistment decisions. Given that Operation Enduring Freedom and Operation Iraqi Freedom (OEF/OIF) represented a national policy change, its effects must be identified from the aggregate time series and thus can be captured in our first but not our second specification. We proxy the effects of the wars in Iraq and Afghanistan by using counts of casualties over time. Given the likely importance of the Iraq War in explaining attitudes toward military service, we also examine the sensitivity of our estimates by using an alternative approach for modeling the effects of the war, described below. Finally, we include the percentage of population in a given state and quarter who are male, black, and Hispanic, since these demographic characteristics may influence taste for military service.

Other Modeling Considerations

Rates Versus Counts

Although some research on enlistment supply focuses on enlistment counts, given the substantial size heterogeneity across different states within the United States, we express our variables as population rates or averages to provide greater comparability across units. Population-weighting the regressions permits us to account for the fact that larger states provide more information about the relative influence of different supply factors.

Logs Versus Levels

Some prior studies estimate enlistment supply models using the raw enlistment rate. In this analysis, we log transform the enlistment rate. For independent variables, log transformation then generates coefficients that are easily interpretable as elasticities and, in some cases, help to correct for a skewed population distribution of the underlying variable. For explanatory variables measuring population proportions (e.g., unemployment rate and veteran status), however, there is no strong reason to prefer logs over levels. For these variables, we chose the transformation that provides the best fit to the data.

Goals

Past research indicates that recruiter effort is an important determinant of recruiting success (Dertouzos, 1985). Insomuch as recruiter effort is correlated with other explanatory variables of interest, failure to control for effort may bias estimates of the relationship between high-quality enlistments and resources or other factors. Unfortunately, effort is not directly observable. Following the control function literature (Heckman and Navarro-Lozano, 2004), we attempt to capture the effects of recruiter effort by including recruitment goals as an additional explanatory variable. Here, we assume that conditioning on goals removes the dependence between ε_{it} and recruiter effort in Equation (3.1). Given that effort may vary nonmonotonically with goals, we flexibly model this relationship using a quartic polynomial in goals.

CHAPTER FOUR

Enlistment Results

In this chapter, we discuss the results of the estimation of the Army and Navy enlistment models and use the results of the models to conduct some "what if" simulations. In particular, we consider "What would have happened to Army and Navy recruiting if bonuses had not been changed between FY 2005 and FY 2008?" We also present marginal cost estimates of bonuses and other recruiting resources.

Estimated Effects of Army and Navy Models

Table 4.1 reports coefficients from regression estimates with heteroskedasticity-robust standard errors clustered at the state level. The table presents the results of the two specifications discussed in the previous chapter for each service.

Specification I for the Army and Navy reports coefficient estimates from our empirical estimation of Equation (3.1). As discussed in the previous chapter, to preserve our ability to estimate the effects of military pay and the Iraq War, both of which vary only at the national level, we do not include a full set of time fixed effects in these specifications. We instead include a polynomial in time to control for general time trends.[1] We reiterate that the regression results for both specifications measure associations between the variables in the model and do not represent causal effects. In particular, because bonuses and other resources are chosen by policymakers in the services, the Office of the Secretary of Defense (OSD), and Congress according to the conditions prevailing in particular locations and times, some of the associations we describe may reflect factors that are unmeasured within the model.

We find that Army recruiters and bonuses enter positively and significantly, with the estimated recruiter elasticity of 0.625, commensurate with elasticities obtained in past studies. The estimated bonus elasticity of 0.055 indicates that a 10 percent increase in the average unconditional bonus amount is associated with a 0.55 percent increase in enlistment contracts. The unemployment rate is also positively and significantly related to high-quality enlistment contracts, with a 1 percentage point increase in the unemployment rate (for example, from 5.5 percent to 6.5 percent) generating a roughly 2 percent increase in contracts.

---

[1] We experimented with different functional forms for the time trends; a quartic polynomial provided the best balance between fit and parsimony. Including higher-degree time trends did not substantially affect the coefficient estimates, but some differences were observed for models including linear or quadratic time trends only. The sensitivity of these results to the parameterization of the time trends underscores our point regarding the difficulty in properly identifying the contributions of factors that vary only over time.

Table 4.1
Coefficient Estimates of Army and Navy Enlistment Supply Models, Dependent Variable = Log(High-Quality Enlistments/Population)

	Explanatory Variable			
	Army		Navy	
	I	II	I	II
Log(recruiters/population)	0.624** (0.0421)	0.570** (0.0579)	0.409** (0.0648)	0.217** (0.0791)
Log(bonus amount)	0.0551** (0.0147)	0.173** (0.0451)	−0.0187 (0.0137)	0.0650** (0.0241)
Percentage receiving college fund	−0.117 (0.107)	−0.127 (0.137)	0.161 (0.131)	−0.0591 (0.144)
Log(unemployment rate)	0.107** (0.0367)	0.101* (0.0429)	0.121** (0.0279)	0.117** (0.0253)
Log(% veteran)	0.325 (0.223)	0.311 (0.221)	−0.241 (0.179)	−0.178 (0.155)
Log(% noncitizen)	−0.0401† (0.0227)	−0.0405† (0.0224)	0.0174 (0.0240)	0.00654 (0.0215)
Percentage enrolled in college	−0.0521 (0.0577)	−0.0493 (0.0571)	0.0531 (0.0544)	0.0349 (0.0531)
Percentage male	0.0173 (0.0304)	0.0223 (0.0300)	0.00887 (0.0293)	0.0276 (0.0275)
Percentage black	0.0544** (0.0205)	0.0598** (0.0220)	0.0389 (0.0329)	0.0480 (0.0333)
Percentage Hispanic	0.00218 (0.00367)	7.31E−4 (0.00384)	0.0185** (0.00556)	−0.00264 (0.00345)
Log(military/civilian wage)	1.15** (0.131)		0.733** (0.182)	
Log(maximum MGIB benefit/tuition)	0.136* (0.0591)		0.0532 (0.0948)	
OEF/OIF casualties	−4.01E−4** (5.28E−5)		−3.68E−4** (7.92E−5)	
No. of observations	1,800	1,800	1,800	1,800
R^2	0.896	0.912	0.832	0.875
Include state fixed effects?	Yes	Yes	Yes	Yes
Include time fixed effects?	No	Yes	No	Yes

NOTES: The table reports coefficient estimates from a regression relating the log number of high–quality enlistments per population to factors affecting enlistment supply. The regression is estimated using state effects only (specification I) or state and time effects (specification II). The unit of observation is a state and quarter; the sample includes the 50 U.S. states during the period between Q4 1999 and Q3 2008. Standard errors clustered on state are reported in parentheses.

* Denotes statistical significance at the 5 percent level.

** Denotes statistical significance at the 1 percent level.

† Denotes statistical significance at the 10 percent level.

Relative to past studies, such as those reviewed in Asch, Hosek, and Warner (2007), our unemployment elasticity estimate of 0.107 is somewhat low, with typical elasticities being closer to 0.3. This discrepancy might reflect the fact that our sample encompasses an era of historically low unemployment, with national unemployment peaking at only 6.3 percent in mid-2003 and remaining below 5 percent over much of our sample period. Somewhat unintuitively, the point estimates for the effects of the Army College Fund are negative, although the confidence intervals for these estimates encompass both positive and negative values of reasonable magnitude, meaning that we are unable to draw strong conclusions regarding the effectiveness of this resource. Increases in the black share of the population are also associated with greater numbers of high-quality contracts, which is unsurprising, given that the Army has traditionally enjoyed success in recruiting minorities. The coefficients on other demographic characteristics of the population, although typically in the expected direction, were generally not statistically significant. Although the effects of recruiter goals are not a primary focus of this study, we also note that our coefficient estimates for goal polynomials indicated that increases in Army goals have a small but statistically significant effect on high-quality enlistments, but only at low goal levels.[2]

The estimated Army pay elasticity is 1.13. Between FY 2005 and FY 2008, military pay grew by about 14 percent in real terms, which, given this fairly substantial elasticity, partially offset other factors that weakened the recruiting environment.

For the Navy (specification I), recruiters but not bonuses are positively and significantly related to high-quality enlistment contracts, whereas increases in the unemployment rate are associated with contract growth. The estimated pay elasticity of 0.736 is below that of the Army and comparable to some past estimates for the Navy, including Ash, Udis, and McNown (1983) and Hogan et al. (1996). Given that the operational tempo and casualty rates of the Navy have remained relatively stable since 9/11, it is perhaps surprising that the estimated negative effect of casualties for the Navy is of comparable magnitude to that for the Army. One possibility is that the casualty measure captures more general attitudes toward the wars in Iraq and Afghanistan. As in the Army, the Navy College Fund benefits were not associated with statistically significant increases in contracts, and Navy goal increases were associated with increased high-quality contracts at moderate and high goal levels but not at the lowest levels.

The table also reports estimates from specification II for each service that include a full set of time fixed effects. These specifications avoid confounding effects that result from unobserved factors affecting the entire country and thus do not permit separate estimates of the effect of Iraq casualties or military pay. These estimates do not exploit the aggregate variation in bonuses depicted in Figures 2.1 and 2.2, instead essentially comparing changes in enlistments in states with large versus small increases in bonuses to identify the effects of bonuses.

Estimated elasticities for Army recruiters, unemployment, and other covariates remain largely unchanged after including time fixed effects on the model. One difference is the estimated bonus elasticity, which increases for both the Army and the Navy. We would expect the estimated bonus elasticity in specification I to be biased downward relative to the true elasticity for each service if bonuses increased at the same time that unobserved national-level factors acted to reduce propensity to join the military. This possibility seems particularly salient given that the most dramatic increases in bonuses at the national level occurred between FY 2005

---

[2]   It is interesting to note that our results also change little if we exclude goals as controls.

and FY 2008, a period when popular support for the war in Iraq moderated. Although our inclusion of goals and time trends may help to reduce such omitted variable bias, the lower bonus elasticities obtained in specification I with less robust time controls than in specification II seem suggestive of the possibility of bias. That is, when we include time fixed effects in specification II to control for such national trends as the change over time in enlistment propensity, we find a higher bonus elasticity than in specification I, where we use less robust controls for national time trends. Navy recruiter elasticity also falls after including time fixed effects.

Our estimated bonus elasticities are above those found in past studies, many of which find elasticities that are not significantly different from zero and in some cases even negative (Asch, Hosek, and Warner, 2007). One interpretation of the data is that bonuses have become more effective recently than they have been historically. For example, in recent years, the Army has heavily advertised the availability of bonuses, which could explain both the increase over time in the market expansion effect of Army bonuses and the higher effect observed in the Army relative to the Navy, which does not aggressively advertise bonuses. Alternatively, it may be the case that the large recent increase in bonuses provides better variation to identify the effects of bonuses than was available for past studies that focused on periods with much more stability in bonus amounts.

To understand the magnitude of the estimated bonus elasticities, consider the fact that bonuses rose from an average of roughly $3,000 to $14,000 for the Army between FY 2004 and FY 2008 (1.5 log points) and $3,000 to $4,500 for the Navy (0.4 log points). With estimated bonus elasticities of 0.17 and 0.07, bonuses can explain an almost 30 percent increase in Army enlistments and a 3 percent increase in Navy enlistments.

The Effects of the Iraq War on Army Enlistments

Table 4.1 demonstrates that casualties in Iraq and Afghanistan are negatively and significantly related to Army enlistments. At their peak, quarterly casualties in Iraq reached approximately 300, which could explain a peak contract decline of roughly 12 percent in our model estimates. However, although casualty data are potentially useful because they represent an objective measure linked to operational conditions, they may understate the effects of the war. For example, survey evidence demonstrates that among blacks, an important constituency for the military, support for the Iraq War declined precipitously *before the onset of ground operations*, and support for the war has been shown to be highly predictive of willingness to recommend military service (Asch, Heaton, and Savych, 2009). Indeed, our estimated contract effects attributable to the war are smaller than previous estimates such as Asch, Heaton, and Savych (2009) and Simon and Warner (2007).

An alternative way to measure the effects of the Iraq War is to attribute to the war all otherwise unexplained national changes in enlistments that occurred after Q1 2003, when ground operations commenced.[3] This procedure is clearly incorrect, given that many other national-level factors that also changed over this period are not explicitly incorporated into the model. In particular, this approach might overstate the negative effects of the war if recent

---

[3] As a practical matter, we accomplish this by estimating a version of our model that excludes casualties but has a set of time dummies for each quarter beginning with Q1 2003. The coefficients on these variables measure the unexplained aggregate time-series variation.

increases in recruiting resources reflect not only a response to the war but also other factors outside the model that negatively affect enlistments. Nevertheless, if the war represents the predominant factor affecting enlistments since FY 2003, then this approach may provide a reasonable approximation of the war's effects. Moreover, comparing the effects obtained using this approach to our estimates of effects based on casualties—which we believe understates the war's effects—provides one potential way to bound the effects of the war.

Figure 4.1 plots the estimated effects of the Iraq War over time obtained using the two methods. To obtain the effects based on casualties, we computed the difference between the predicted number of quarterly contracts using specification I and the predicted number of contracts using specification I but assuming zero casualties. Similarly, to obtain the time-series effects, we calculated the difference between the predicted values from specification II with time fixed effects and the predicted values assuming time fixed effects of zero beginning in FY 2003.[4] For ease of interpretation, we have translated the predicted effects into percentage changes by dividing by the mean number of quarterly contracts in the prewar period between FY 2000 and FY 2002. Effects estimates for the casualty-based method generally range between 0 percent and –10 percent, with an average effect of –6 percent over the entire time period. Given that casualties diminished somewhat in FY 2008, our corresponding estimates of the effect of the war also fell in this year to –2 percent to –3 percent.

The time-series approach suggests a much larger effect of the war, with a short-lived positive bump generated by the successful ground invasion followed by increasingly acute negative effects. By FY 2006, the time-series method suggests that the war could account for a 50 to 60

Figure 4.1
Predicted Percentage Change in High-Quality Army Contracts Resulting from the Iraq War, Using Alternative Estimation Methods

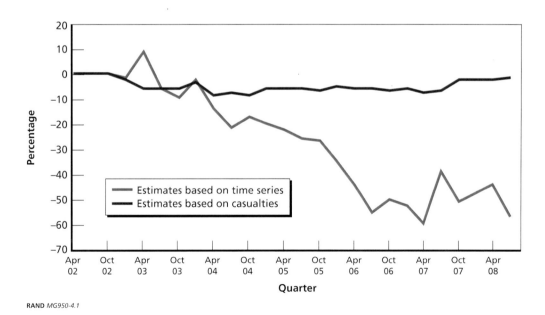

---

[4] Conceptually, it might make most sense to use the actual number of contracts as the baseline for comparison. As a practical matter, because our dependent variable is disaggregated across states and log transformed, there are minor differences between the predicted and the actual total quarterly contracts, even when we include time fixed effects.

percent decline in contracts, a substantial effect. It is interesting to note that this approach also suggests that negative effects of the war have stabilized since mid-2006.

The considerable variation in effects estimates yielded by the two approaches highlights the difficulties inherent in closely measuring the effects of national policy changes in an aggregate model. Nonetheless, the approaches are consistent in indicating that the war had a sizable negative effect on high-quality enlistments, even though the precise magnitude of the effect remains somewhat uncertain.

Simulations of Policy Scenarios

One simple way to characterize the effects of the large increase in bonuses that occurred between FY 2005 and FY 2008 is to use the model to hold average bonus levels fixed at the amounts observed in FY 2004 and to project high-quality enlistments under this alternative policy scenario. In other words, we use the model to simulate what would have happened to Army high-quality enlistments had bonuses remained fixed rather than increased over this period.

Figure 4.2 plots actual high-quality enlistments and projected high-quality enlistments without the bonus increase for the Army. Projections are based on the specifications that include both time and state fixed effects.

Our model estimates indicate that, between October 2004 and September 2008, the enlistment bonus expansion was associated with an average increase in Army high-quality enlistments of 1,669 contracts per quarter. Over the entire period, the bonus expansion yielded

Figure 4.2
Actual Army High-Quality Enlistments and Simulated Enlistments in the Absence of an Increase in Enlistment Bonuses

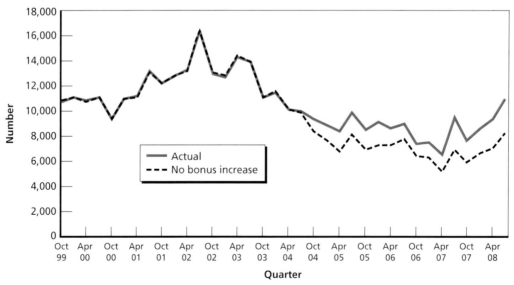

an additional 26,700 high-quality contracts, or 20 percent of the total number of high-quality enlistment contracts obtained.[5] Bonuses can account for over 6,000 additional high-quality contracts in each fiscal year between FY 2006 and FY 2008, a period in which the Army exceeded its annual overall recruiting goal by fewer than 1,000 recruits. Although the bonus expansion came at significant financial cost to the Army, our estimates suggest that the Army would have been unable to meet its mission had bonuses remained at FY 2004 levels absent other potentially costly resource expansions.

Given the smaller increase in bonuses for the Navy and smaller estimated bonus elasticity, we would expect more modest projected effects for the Navy. Indeed, as Figure 4.3 demonstrates, the projected effect of the Navy bonus expansion on enlistments is minimal, with a fairly similar trajectory of enlistments predicted in the absence of a bonus increase. On average, the Navy's more generous bonuses produced an additional 109 high-quality contracts per quarter, or 1,700 contracts over the entire period. This finding is consistent with the within-occupation bonus patterns we documented in the previous chapter, which suggested that despite an increase in absolute bonus amounts, bonuses have continued to be used primarily as a skill-channeling device in the Navy.

Figure 4.3
Actual Navy High-Quality Enlistments and Simulated Enlistments in the Absence of an Increase in Enlistment Bonuses

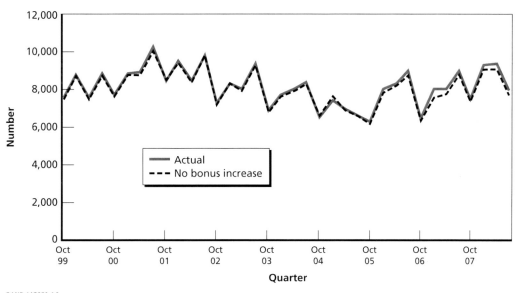

RAND *MG950-4.3*

_____

[5] These projections do not take into account the uncertainty associated with our elasticity estimates. However, given that the 95 percent confidence interval for the bonus elasticity excludes values below 0.084, even using more conservative values for the bonus elasticity that are consistent with the estimates would indicate that bonuses generated a sizable number of enlistments.

Estimates of Marginal Cost for the Army

By combining the above estimates of the increase in enlistments associated with changes in various recruiting resources with information about the costs of these resources, we can examine the cost-effectiveness of bonuses relative to other resources. We are particularly interested in understanding whether the large expansion in bonuses that occurred between FY 2005 and FY 2008 was cost-effective in the sense that the cost of generating the additional 26,700 high-quality contracts between FY 2005 and FY 2008 estimated above by means of bonuses was less, per recruit, than the cost of generating a similar number of high-quality recruits using other types of resources. Therefore, we first compute the cost of the bonus increase. We then compute the amount by which other resources would have had to increase to generate the same 26,700 increase in the number of in high-quality recruits, again using our estimated model to make this computation. Finally, we compute the cost of these increases in other resources and compare it to the cost of bonuses. We compute marginal costs, or the change in cost associated with generating an additional recruit.

We emphasize from the outset that these cost estimates are affected by numerous sources of uncertainty. In addition to the statistical uncertainty associated with our elasticity estimates, our model encapsulates a specific set of assumptions regarding the factors affecting bonuses that are not directly testable. As explained above, the limitations inherent in our modeling approach make us cautious in interpreting the estimates as reflecting causal relationships, a necessary ingredient to producing cost comparisons. Moreover, our cost data themselves provide only noisy estimates of the precise costs of different enlistment resources and abstract from such issues as scale effects.[6] Thus, the cost comparisons are best interpreted as providing a general indication of the relative cost effectiveness of different resources. We also focus on the effects of each incentive on altering the number of recruits rather than the number of recruit years. These quantities could differ if incentives affect not only the number of new recruits but also their length of service. Unfortunately, our aggregate model is not well suited to estimating any length-of-service effects.[7] We discuss below the specific calculations and assumptions underlying our cost numbers for each resource.[8]

Cost comparisons, although informative, also obscure important differences in the nature of recruiting incentives that should be included in any policy discussion of the optimal mix of incentives. Pay increases, for example, are generally implemented across all services and all pay grades, making it difficult to target these incentives to services experiencing shortfalls. In addition, because pay changes must be implemented legislatively and default pay increases are

---

[6] Scale effects occur when the cost of providing an additional amount of a resource changes as the total amount spent on the resource varies. For example, once there is an administrative and management system in place for selecting and assigning recruiters, it may be fairly easy to add an additional recruiter. However, such systems are unlikely to be developed until the total recruiter force size reaches a certain level.

[7] In the next chapter, we present a way to estimate length-of-service effects for bonuses using individual data.

[8] In the discussion that follows, we focus on the financial costs to the services of increasing particular resources to obtain more high-quality recruits. Other cost concepts have been used in prior literature and the use of these concepts might affect the conclusions. For example, bonus expansions may involve giving increased bonuses to some individuals who would have signed contracts at lower bonus levels. In considering the financial costs of bonuses, it is appropriate to include such expenditures as costs. However, under other cost concepts, such as "social cost," which would recognize the fact that the recipients of such payments obtain rents, which are valuable to them, such payments might be excluded. Similarly, in the treatment of pay, the handling of rents received by lower-quality recruits will affect the cost calculations. Thus, bonuses and pay may appear more costly when considering financial costs than they do under alternative cost metrics.

defined by a statutory formula, pay increases cannot be used to respond to short-run fluctuations in the recruiting environment.

Similarly, recruiters provide considerable flexibility to focus on particular geographic areas. Recruiters' time can also be used for "investment" activities designed to hedge against future downturns in the recruiting environment.[9] However, rapidly changing the size of the recruiter force is difficult, because recruiters require training and some less easily adjusted inputs, such as office space. Additionally, because recruiters are often drawn from other occupational specialties, altering the number of recruiters can have downstream career-management effects that are not present for other types of resources.

Relative to pay and recruiters, bonuses can be more directly targeted at particular occupations, and bonuses can also be changed relatively quickly to respond to short-run developments in the recruiting environment. However, bonuses are potentially more likely than other incentives to generate "skimming effects," whereby bonuses offered in one service attract recruits who may have otherwise joined a different service. Therefore, although cost-effectiveness is one criterion for comparing recruiting resources, other considerations may also be important.

Bonuses

Table 4.2 reports the aggregate expenditures on Army enlistment bonuses by fiscal year calculated from accounting data reported by DoD in the annual military personnel budget document for the Army (U.S. Department of the Army, various years). Although the accounting data likely provide the most accurate portrait of the overall financial cost of bonuses to the services, because bonuses are paid out over time, bonus expenditures in one fiscal year partially reflect investments made to attract enlistments in prior years, making it challenging to properly isolate expenditures associated with a particular year of contracts.

Table 4.2
Army Annual Enlistment Bonus Expenditures
Using Accounting Data, FY 2000–2008

Fiscal Year	Expenditures ($)
2000	119.2
2001	202.3
2002	240.6
2003	176.0
2004	226.7
2005	183.8
2006	377.4
2007	493.7
2008	422.3

SOURCE: Department of the Army, various years.
NOTE: Amounts are in millions of FY 2008 dollars.

---

[9] For example, recruiters can spend time on relationship-building activities, such as meeting with guidance counselors, thereby investing in their ability to identify potential future recruits.

However, the Army also provided RAND with contract data that report bonus amounts associated with each enlistment contract, permitting us to compare aggregate payment amounts in the contract data to those in the accounting data to obtain a better sense of the correct timing for bonus expenditures. As demonstrated in Figure 4.4, after adjusting for the fact that some bonus costs recorded in the contract data are ultimately not realized by the Army because of early separations, expenditures recorded in the accounting data reported in Table 4.2 closely track those promised in the prior fiscal year of the contract data.[10] For example, from Table 4.2 we expect the $184 million in bonus payments in FY 2005, for example, to largely reflect bonus commitments made in the prior year.

To approximate the cost of the bonus increase, we assume that the actual cost of enlistment bonuses associated with FY 2004 contracts was $184 million and that any increases in bonus expenditures in subsequent years reflect the expansion. Relative to the $184 million baseline level of bonuses, over the next three years, additional bonus expenditures totaled $742 million, or $247 million per year.[11] Although actual bonus expenditures for FY 2009 (which largely reflect FY 2008 contracts) are not yet available, we project, using our contract data, that actual bonus expenditures associated with FY 2008 contracts would be $639 million, or an increase of $455 million relative to the baseline, placing the total cost of the four-year bonus expansion at $1.20 billion ($742 + ($639 - $184) = $1,197 million), or $300 million per year. In light of our analysis above indicating that the bonus expansion generated 26,700 additional high-quality recruits, the cost per recruit generated was $44,900 ($1.2 billion divided by 26,700).

Figure 4.4
Actual and Contracted Enlistment Bonus Payments for the Army (in FY 2008 dollars)

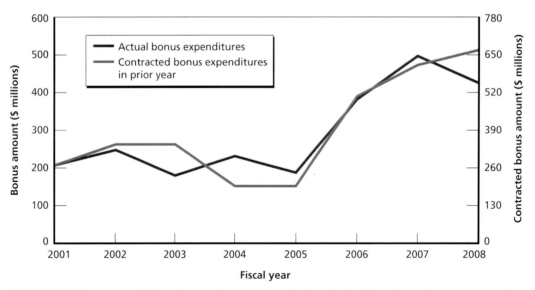

SOURCES: U.S. Department of the Army, various years, and authors' calculations from Army contract data.
RAND MG950-4.4

[10] In particular, in each year, the promised payments in the contract data are 30 percent higher than the actual payments made the next year.

[11] The calculation for the increase is (377.4 – 183.8) + (493.7 – 183.8) + (422.3 – 183.8) = $742 million.

Recruiters

To compare the cost-effectiveness of bonuses with that of recruiters, we use our model estimates to project the number of additional recruiters that would be required to generate 26,700 additional high-quality recruits over a four-year period. As indicated above, our model projects that the $1.2 billion expansion in bonuses generated 26,700 high-quality recruits.

Given our recruiter elasticity estimate of 0.57 in Table 4.1, and the distribution of recruiters across states, our model indicates that a 45 percent increase in the number of recruiters would have been required to keep overall enlistments at observed levels between FY 2004 and FY 2008 had no bonus increase occurred.[12]

Table 4.3 reports the expenditures on Army recruiters computed from actuarial data provided to RAND by the Directorate of Accession Policy within the Office of the Secretary of Defense along with the total number of Army recruiters by fiscal year.[13] Expenditures include recruiter compensation and recruiter support costs.[14] The table also reports the required increase in the recruiter force needed to produce 26,700 contracts and estimates the costs of these additional recruiters by multiplying by the average cost per recruiter.

Table 4.3
Projections of the Additional Cost of Army Recruiters, FY 2000–2008

Fiscal Year	Total No. of Recruiters	Actual Expenditures on Recruiters ($)	Projected Additional Recruiters (45 Percent Increase)	Projected Additional Costs ($)
2000	5,880	502.7		
2001	6,110	507.9		
2002	5,838	580.9		
2003	5,508	603.5		
2004	4,552	584.4		
2005	5,180	493.0	2,331	221.8
2006	5,982	501.3	2,692	225.6
2007	6,109	542.3	2,749	244.0
2008	6,476	501.7	2,914	225.7

SOURCES: Office of the Secretary of Defense and authors' calculations.
NOTE: Amounts are in millions of FY 2008 dollars.

[12] In the calculations that follow, we assume that the effect of recruiters and the marginal cost of a recruiter do not change substantially with the size of the recruiter force. However, given that a 45 percent increase in the number of recruiters leads to a much larger recruiter force than has actually been observed in recent years, such an extrapolation requires considerable caution.

[13] Because the number of recruiters fluctuates from month to month, these are averages over each 12-month cycle. As discussed in Chapter Three, the data on recruiters were provided to RAND by the Army.

[14] Our pay data actually include the total pay costs of military personnel involved in recruiting. The bulk of these are recruiters, but some nonrecruiters are included in these totals. On the other hand, we do not include the cost of some civilians who support recruiters, since we observe only the total cost of civilians involved with recruiting, and many civilians are involved with other programs, such as incentive programs.

Summing the figures in the final column of Table 4.3, the total projected cost of obtaining a similar number of high-quality contracts using recruiters as enlistment bonuses is $917.1 million, or $230 million per year. On a per recruit basis, the cost of a recruiter-based policy is $33,200 per recruit ($917.1 million divided by 26,700). Consistent with past research identifying recruiters as a particularly cost-effective resource, recruiters appear less costly than bonuses.

Pay

As discussed above, the additional enlistment bonuses increased high-quality contracts by approximately 20 percent between FY 2004 and FY 2008. Using our pay elasticity estimate of 1.15, increasing high-quality contracts by 20 percent by increasing military pay would require a roughly 20 percent increase in real military compensation. Real regular military compensation (RMC) averaged $29,346 in FY 2008 dollars over the four-year period in question, and our model indicates that a pay increase of $6,100 would yield a comparable number of recruits over this period. We reemphasize that these calculations are sensitive to the assumed value of the pay elasticity, and that our elasticity is estimated using only aggregate national changes, which is not an ideal source of variation.

Given that changes in regular military compensation are not targeted solely at particular individuals in particular years, but instead affect the entire pay table, we must make admittedly arbitrary decisions regarding whose pay to include in the cost calculations. We follow the convention in the literature and include the costs of regular military compensation for all new recruits, not simply high-quality recruits, but include only RMC costs incurred during the first year of service.

Table 4.4 summarizes the projected costs of an increase of 26,700 high-quality recruits over a four-year period, obtained solely through adjustments to RMC. The total four-year cost of an RMC-based policy would be $1.54 billion, a magnitude 28 percent above the four-year cost of enlistment bonuses. On a per contract basis, the cost of RMC is $57,600.

Advertising

Data limitations force us to exclude advertising from our analysis, but it is constructive to compare our marginal cost estimates to previously published estimates of the marginal cost of advertising. Dertouzos (2009) estimates that the Army was able to generate a 6.7 percent increase in contracts each year in FY 2002 to FY 2003 through advertising, a figure that includes both high- and low-quality contracts. Annual advertising expenditures were roughly

Table 4.4
Total Cost of a Change in Regular Military Compensation for New Recruits, FY 2005–2008

Fiscal Year	Total No. of Accessions	Cost per Accession ($)	Total Cost ($ millions)
2005	58,273	6,100	355.5
2006	67,306	6,100	410.6
2007	61,434	6,100	374.7
2008	64,902	6,100	395.9

SOURCE: Authors' calculations from Army contract data.

NOTE: Amounts are in FY 2008 dollars.

$70 million in these years, in FY 2008 dollars. Assuming a symmetric effect for high- and low-quality recruits, advertising would have generated approximately 3,600 additional high-quality contracts each year, yielding an average advertising cost per new high-quality contract of $19,400. At the same time, given that the advertising estimates were constructed using a different estimation methodology and time period, we are clearly handicapped in this study in our ability to draw conclusions regarding the relative merits of advertising versus other enlistment incentives.

Table 4.5 summarizes our estimates of the additional cost of various enlistment resources per recruit generated, focusing on hypothetical resource expansions implemented during FY 2004–2008. We do not view these as precise estimates, but they do provide general indications of the cost-effectiveness of each resource in terms of market expansion. Enlistment bonuses appear slightly less cost-effective than recruiters but more cost-effective than pay.

Estimates of Marginal Cost for the Navy

Unlike the Army, average bonuses increased fairly modestly in the Navy between FY 2004 and FY 2008. As shown in Chapter Two, although the size of bonuses was rising in real terms for those offered bonuses, a declining share of new applicants received bonus offers. Earlier in this chapter, we demonstrated that the actual bonus changes between FY 2004 and FY 2008 in the Navy had an insubstantial estimated effect on the overall number of high-quality enlistments. Thus, rather than focus on the FY 2004–2008 period for the Navy cost analysis, we instead consider generic changes in the amount of bonuses, number of recruiters, and amount of regular military compensation.

To compute the marginal cost of each resource for the Navy, we use the estimated elasticities in Table 4.1 for each resource. We then calculate the increase in each resource required to achieve a fixed additional number of contracts. For example, with the elasticity estimate for bonuses of 0.065 in Table 4.1, a 1 percent increase in the average enlistment bonus would increase enlistments by 0.065 percent. Across recent years, there have been about 32,000 high-quality Navy enlistments per year, meaning that a 0.065 percent increase would yield approximately 21 additional enlistment contracts. Similarly, our recruiter elasticity estimates in Table 4.1 imply that obtaining 21 additional contracts using only recruiters would require an increase in the recruiter force of 0.16 percent. To yield 21 contracts using pay, pay would have to increase by only 0.09 percent. Thus, a 1 percent increase in bonuses, a 0.16 percent increase in recruiters, and a 0.09 percent increase in pay are approximately equivalent in terms of their effect on high-quality enlistments.

The next step is to compute the additional cost associated with increasing high-quality Navy enlistments by 21 contracts, using each type of resource. Table 4.6 reports annual and average expenditures by the Navy for recruiters and enlistment bonuses using accounting data

Table 4.5
Estimated Marginal Cost of Recruiting Resources, Army

Resource	Bonuses	Recruiters	Pay
Cost per recruit	44,900	33,200	57,600

SOURCE: Authors' calculations.

NOTE: Amounts are in FY 2008 dollars.

Table 4.6
Annual and Average Expenditures for Recruiters, Enlistment Bonuses, and RMC, Navy, FY 2005–2008

Fiscal Year	Recruiters	Bonuses	Pay
2005	482.5	143.1	1,046.2
2006	437.8	207.2	1,042.4
2007	435.1	205.9	1,054.1
2008	410.7	192.2*	1,091.9
Average	441.5	187.1	1,058.6

SOURCES: Office of the Secretary of Defense and authors' calculations from Navy contract data.

NOTE: Amounts are in millions of FY 2008 dollars.

* Denotes a projected cost.

provided by the Directorate of Accession Policy within OSD. Following our approach for the Army, we assign to each fiscal year the bonus expenditures from the preceding year. To estimate annual Navy expenditures for pay, we multiply the annual number of accessions by the real value of regular military compensation.

Using the average cost figures in Table 4.6 allows us to calculate the cost per Navy high-quality enlistment for bonuses, recruiters, and pay. The cost per Navy high-quality recruit is reported in Table 4.7.

Navy enlistment bonuses are less cost-effective at market expansion than recruiters or pay. This finding is expected, given the relatively low estimated bonus elasticity for the Navy, reflecting the fact noted above that the Navy continued to use bonuses during this period primarily for skill-channeling rather than for market expansion purposes. Although the recruiter cost per high-quality recruit is approximately equal across the two services, bonus costs are higher in the Navy than in the Army.

Concluding Thoughts

The analysis in this chapter indicates that 20 percent of the increase in high-quality enlistments between FY 2004 and FY 2008 can be explained by increases in enlistment bonuses. Our simulations suggest that had bonuses remained at the levels observed in FY 2004, the Army would have attracted substantially fewer recruits, jeopardizing recruiting goals during a period when the conflicts in Iraq and Afghanistan contributed to a very challenging recruiting environment for the Army. Cost-effectiveness estimates for bonuses are in line with those

Table 4.7
Estimated Marginal Cost of Recruiting Resources, Navy

	Bonuses	Recruiters	Pay
Cost per recruit	89,100	33,600	45,400

SOURCE: Authors' calculations.

NOTE: Amounts are in FY 2008 dollars.

for other major market expansion tools, such as recruiters, and are somewhat lower than pay. The Navy results suggest that bonuses played a more traditional role over this period, to skill-channel recruits rather than expand the market. We find that as a market expander, enlistment bonuses are not cost-effective for the Navy, validating the Navy's decision to hold stable their average value.

CHAPTER FIVE

Army Attrition Results

The aggregate analysis presented in Chapter Four is designed to estimate the relationship between the availability of enlistment bonuses and the number of high-quality enlistments. Enlistment bonuses may have effects beyond the simple enlistment decision and the occupational choice decision; they might affect the likelihood that individuals who enlist fulfill their enlistment term or leave before completing their enlistments. From a policy standpoint, understanding the relationship between both bonuses and enlistment and bonuses and first-term attrition is important to properly understand how policies that affect enlistment bonuses are likely to affect the number of recruits and the average number of person-years during the first term provided by each recruit.

Enlistment bonuses may affect attrition through two channels. First, enlistment bonuses affect the incentive for early departure, because in cases where bonuses exceed a threshold amount, they are paid in installments over the first term of service. Individuals who terminate their military service before the end of their term are ineligible to receive outstanding bonus payments.[1] Thus, recruits receiving more than the threshold amount have a financial incentive to fulfill their service obligation. Second, enlistment bonuses may also induce individuals with an otherwise lower-than-average taste for military service to enlist. If such individuals are more likely to attrit during their first term, then this selection effect of bonuses would tend to generate a positive relationship between enlistment bonus receipt and first-term attrition.[2]

To estimate the relationship between bonus receipt and attrition for the Army, we turn to an individual-level analysis. For each individual who accessed on or after 10/1/1999 and whose required term of service ended before October 1, 2008, we construct an indicator variable for whether the individual completed the first term of service using the individual's accession date, enlistment term, and separation date.[3] We then estimate probit regression equations of the following form:

$$Y_i = \alpha \times Bonus_i + \beta X_i + \varepsilon_i \qquad (5.1)$$

The variable Y is the propensity to separate before the end of the first term. We do not observe propensity; we observe only whether the individual left service before the end of the

---

[1] Military regulations also require that individuals who fail to complete a contracted term of service reimburse the government for portions of their enlistment bonus.

[2] This effect is not unique to enlistment bonuses, but it can occur for any incentive designed to increase enlistments.

[3] We code individuals who separated less then 60 days before the end of their term of service as having completed their required service.

first term. We estimate a probit model of the probability that the Army recruit left before the end of the first term as a function of enlistment bonus receipt and a set of other controls, X_i. Included in X_i are controls for gender, race, marital status, number of dependents at the time of enlistment, waiver status, and AFQT and fixed effects for age, year of enlistment, enlistment term, educational attainment, and MOS. AFQT is entered quadratically to allow for the possibility that individuals with particularly high or low aptitude may be more likely to attrit. The coefficient α measures the expected difference in the probability of attrition for individuals who did and did not receive an enlistment bonus who have otherwise comparable demographic characteristics.

If enlistment bonuses were randomly assigned to potential recruits, then α would measure the causal effect of bonuses on attrition, not just the association between bonuses and attrition. However, bonuses are not randomly assigned but are targeted at individuals who are deemed more desirable enlistments, either because of superior education and skills or because of a willingness to assume certain hard-to-fill assignments. Such targeting of bonuses is expected to induce a correlation between the receipt of bonuses and the error term in Equation (5.1), creating what is known as an endogeneity problem. Because of targeting, the receipt of bonuses may depend on characteristics that affect attrition. Estimates of α that do not account for this correlation will confound the effect of bonus receipt with the effects of other enlistee characteristics that affect attrition and that also affect bonus receipt.

We note that targeting and the resulting econometric problem of selection bias are not problems in this context, as they are in the context of many other labor market studies. Indeed, an important function of enlistment bonuses is to select particular types of individuals into the military. Thus, estimates of the causal effects of bonuses will of necessity incorporate such selection effects. Instead, the endogeneity problem arises because there may be omitted characteristics such as noncognitive skills that affect attrition and bonus receipt that we would wish to control for but cannot observe.

To estimate the causal effect, we require a factor or "instrument" that affects bonuses but that is exogenous with respect to other characteristics of enlistees. Our instrument for bonus receipt is the designation of an MOS as critical at the time of enlistment. At any particular time, certain MOSs are designated as critical by Army leaders as a result of projected personnel needs. Critical occupations are more likely to be eligible for enlistment bonuses. If the timing of when particular occupations are deemed critical is not correlated with the characteristics of enlistees, critical status provides a potential source of exogenous variation in bonuses that can be used to identify the effects of bonuses on attrition. Because the analysis includes a full set of MOS indictors or fixed effects as control variables, the effect of bonus receipt is identified by comparing the attrition rates of individuals who enlisted into the same occupation but who differed in whether they received a bonus based on the fact that some individuals enlisted during periods in which the occupation was deemed critical.[4]

For critical status to be a valid instrument, designation of an MOS as critical must be uncorrelated with unobservable characteristics of enlistees conditional on the observed char-

---

[4] A related approach that might better resolve selection problems would be to use a regression discontinuity (RD) design that compares the outcomes of those who enlist just before or just after the date at which new administrative rules are introduced that change the level of available bonuses. A drawback of the RD approach is that it would be MOS-specific and would require a fairly high number of recruits on either side of the cutoff, which would limit that analysis to only those occupations with substantial numbers of enlistees.

acteristics in X_i that are included in the model. We assume this to be the case, but we cannot directly test this assumption empirically to determine its validity. However, we can compare the observable characteristics of enlistees in critical and noncritical MOSs to assess whether the instrument appears to achieve balance on these observed characteristics.

Table 5.1 reports the average demographic characteristics of enlistees in critical and noncritical MOSs conditional on the MOS they choose and their month of enlistment. The sample includes 332,104 individuals and encompasses both high-quality and other recruits.[5] Although some of the differences are statistically significant, which is unsurprising given our large sample size, in practical terms the two populations of enlistees appear fairly similar. The primary differences are a larger proportion of African-Americans and a higher average AFQT score among those enlisting during critical periods.

Table 5.2 reports our results using a standard probit model (column 1) and instrumental variables (column 2). The table shows the estimated effect of bonus receipt on the probability of attrition as well as the estimated coefficients of selected variables in X_i. The estimated coefficient estimates have been transformed into marginal effects at the means of the variables in the model. Given that the endogenous variable (bonus receipt) is binary, we obtain the instrumental variable estimates using a bivariate probit model. The bivariate probit jointly models the distribution of the error term in Equation (5.1) and the error term in a second equation describing the probability of bonus receipt. In our model, we identify the effect of bonuses by including a 0-1 variable measuring a critical MOS designation as an explanatory variable in the bonus receipt equation while excluding it from the attrition equation; both equations include the same set of demographic controls listed above. The results of the bivariate probit model can be used to construct a test statistic that allows us to test for nonzero covariance between the two errors, which would indicate that bonuses are endogenous. In this test, we reject the null hypothesis of exogeneity ($\chi^2(1) = 10.98$, $p < 0.001$), suggesting that the instrumental variable model is preferred.

Table 5.1
Army Enlistment Demographics, by Average Enlistees' Critical Bonus Status

Characteristic	Noncritical Status	Critical Status	T–Test for Difference
Years of education	12.14	12.13	–2.66
Married	0.144	0.149	1.88
Male	0.813	0.833	8.27
Black	0.175	0.154	–7.52
Hispanic	0.095	0.090	–1.84
Number of dependents	0.297	0.308	2.08
Had enlistment waiver	0.137	0.135	–0.69
Age	20.73	20.73	–0.32
AFQT	58.77	61.48	26.69

---

[5] We obtain similar results if we limit the analysis to high-quality recruits only.

Table 5.2
Estimated Effect on the Probability of First-Term Attrition, Army

Explanatory Variable	Standard Probit Model	Instrumental Variable Bivariate Probit Model
Received enlistment bonus	0.0069**	−0.0170*
	(0.0025)	(0.0077)
Married	0.0092*	0.0092*
	(0.0043)	(0.0043)
Male	−0.2360**	−0.2352**
	(0.0026)	(0.0026)
Black	−0.0255**	−0.0254**
	(0.0026)	(0.0027)
Hispanic	−0.0886**	−0.0889**
	(0.0031)	(0.0033)
Additional dependent	0.0155**	0.0156**
	(0.0020)	(0.0020)
Received enlistment waiver	0.0219**	0.0221**
	(0.0029)	(0.0029)
No. of observations	329,358	329,361

SOURCE: Authors' calculations from Army contract data.

* Denotes statistical significance at the 5 percent level.

** Denotes statistical significance at the 1 percent level.

Using the standard probit model approach, the estimated effect of bonuses is positive and statistically significant. Relative to the average attrition rate of 32 percent in our data, the estimated marginal effect of 0.0069 percentage points implies that bonuses increase attrition by approximately 2 percent (0.0069/0.32)—a modest effect. The estimated coefficients on the other covariates are generally statistically significant and accord with intuition—males and minorities are less likely to attrite whereas those who are married, have dependents, or enlist with a waiver are more likely to attrit. It is interesting to note that attrition is only about 7 percent higher among those receiving waivers after controlling for other demographic characteristics, suggesting that waivers do not draw in particularly unqualified recruits.

The results in column 2 versus 1 paint a contrasting picture of the effects of bonuses. After accounting for endogeneity that results from the correlation between unobserved variables affecting attrition and bonus receipt, the effects of enlistment bonuses are now negative and statistically significant. Enlistment bonuses decrease attrition by 5 percent. The estimated coefficients on other covariates are fairly similar in the instrumental variables (IV) specification. The fact that the estimated effect of bonus receipt decreases when we account for possible endogeneity suggests that bonus receipt is correlated with unobserved factors leading to attrition. Such a correlation might occur if, for example, individuals with superior interpersonal skills are more likely to receive bonuses and also have better outside options, resulting in a higher likelihood of leaving the military early. Overall, our analysis demonstrates that beyond an increase in high-quality enlistments, enlistment bonuses have the additional effect of reducing attrition. Thus, the overall effect of enlistment bonuses on person-years is larger than would be implied by an analysis focusing solely on the enlistment decision.

CHAPTER SIX

Background on the Army's Selective Reenlistment Bonus Program

The military services have long used enlistment bonuses and reenlistment bonuses to help attract and retain personnel in critical skills.[1] In FY 1974, soon after the advent of the All Volunteer Force, Congress amended existing legislation governing military reenlistment bonuses to create the selective reenlistment bonus. Since then, the SRB has been the Army's primary tool for managing enlisted retention. This chapter briefly discusses the theoretical basis for such a program and answers in general terms the question of why a bonus program is needed. It first describes the evolution of the Army's SRB program over the period FY 1974–2007 and the significant changes that were made to the program in FY 2007. The chapter then describes the program as it now exists and provides detailed information about the program for FY 2008. A theme of this discussion is that the sophistication with which the Army implements this program has increased over time. Nevertheless, questions about its program remain.

Reenlistment Bonus Program Overview, FY 1974–2007

Over the long course of the SRB program, military reenlistees have received selective reenlistment bonuses determined by the formula SRB = AOS × SRBM × MBP. In this formula, AOS denotes additional obligated service, measured in years.[2] From FY 1974 until mid-2007, Army reenlistees had to obligate for at least three additional years to qualify for an SRB. MBP denotes the service member's monthly basic pay at the time of reenlistment, and SRBM denotes the selective reenlistment bonus multiplier. The Army attempts to influence reenlistments by manipulating the multiplier. SRB amounts are determined partly by soldiers' choices about the length of reenlistment or AOS.

The Army's goal is to set the SRB multiplier at a level that balances its requirement for experienced personnel with the supply of experienced personnel. Requirements vary along many dimensions, including (but not limited to) MOS, rank, and seniority level. Over much of the period since FY 1974, the Army manipulated the SRB multiplier mainly by MOS and reenlistment zone. There are three seniority zones for SRB: Zone A (2–6 years of service), Zone B (7–10 years of service), and Zone C (11–14 years of service). In the mid-1990s, the Army began to set SRB multipliers by rank as well as by MOS and zone. When reenlistments in a particular MOS, rank, and zone cell fall short of requirements for personnel in that cell, the

---

[1] See DoD (2005), pp. 610–631, for a history of legislation governing military reenlistment bonus programs.

[2] AOS is also referred to as length of reenlistment or LOR. In Chapter Seven, we use the term LOR.

Army raises the multiplier. Likewise, when reenlistments in a particular cell exceed requirements, the multiplier is reduced.

Historically, Army SRB multipliers have ranged from 0 to 6 and varied in half-unit increments.[3] A multiplier of 0 indicates that the Army is achieving (or exceeding) its retention goal without the need for a reenlistment bonus. Each half-multiple increase in the multiplier indicates one-half an additional month of basic pay per year over the period of the reenlistment (but based on MBP at the time of reenlistment).

Setting bonus multipliers by MOS, rank, and zone meant that imbalances between requirements (personnel demand) and personnel inventories (personnel supply) might occur at more disaggregated (detailed) levels. One such level of disaggregation is location of assignment. Some geographic locations are less desirable than others, and there might be a shortage of personnel willing to go to such locations without additional compensation even when a category is in balance overall. Furthermore, some subcomponents of military skill might be in short supply even when there is no overall shortage of experienced personnel in the MOS. When a shortfall occurs at a more disaggregated level than one defined by MOS, rank, and zone, it is more efficient to vary the bonus along these extra dimensions than to set a common multiplier for everyone.

To permit SRB management at more detailed levels, in FY 1999 the Army introduced the targeted SRB (TSRB) program. Now SRB multipliers began to be varied by location of assignment and by skill within an MOS. Personnel assigned to Europe and Korea, for example, began to receive larger bonuses than otherwise similar personnel assigned to most bases within the continental United States (CONUS). Within CONUS, personnel assigned to Fort Drum began to receive larger SRBs than otherwise similar personnel assigned elsewhere within CONUS. The TSRB program also provided personnel possessing certain Special Qualification Identifiers (SQIs) larger multipliers than other personnel in the parent MOS. SQIs include P (parachute qualified), G (Ranger), V (Airborne Ranger), and T (Special Forces).[4] Over time, the Army's SRB program has evolved from a relatively simple one with few dimensions to a complicated one with many dimensions.

The conflicts in Afghanistan and Iraq brought additional wrinkles to the SRB program. In FY 2004, the Army introduced the location SRB (LSRB). Like the TSRB, the LSRB program established different (larger) multipliers for personnel in specific units and MOSs that were mobilized to Afghanistan and Iraq than were available to similar personnel in other units. Affected units tended to be combat units that were bearing the brunt of the fighting in Iraq and Afghanistan.

Not all deployed personnel were eligible for LSRBs. So in September 2003, the Army began offering deployed SRBs (DSRBs) to personnel who reenlisted while deployed to Afghanistan, Iraq, or Kuwait and did not otherwise qualify for an SRB. The DSRB began as a lump-sum bonus of $5,000, but the Army quickly converted it to a multiplier-based bonus (in December 2003). In September 2004, the Army settled on DSRB multiplier values of 1.5 for

---

[3] Legislation establishing the SRB program originally permitted SRB multipliers from 0 to 6. The law was amended in FY 1989 to permit multipliers of up to 10. See DoD (2005), p. 625. The Navy is the only service to have taken advantage of this increase in the maximum multiplier.

[4] To be specific, personnel receive the Special Qualification Identifier of T after completing the first Special Forces Operational Detachment–Delta Unit Assessment and Selection Course and receiving 18 months of on-the-job training in a Delta force operational element.

Zone A reenlistments, 1.0 for Zone B reenlistments, and 0.5 for Zone C reenlistments. These DSRB multipliers remained in effect until June 2007.

Whenever the Army makes a change to its SRB program, the changes are announced in "Milpers" messages issued by the Army Human Resources Command. Over the period FY 1997–2007, the Army issued 125 messages relating to the SRB program. Especially since the beginning of operations in Afghanistan and Iraq, the changes became more frequent. In FY 1997, the Army issued four SRB-related messages. In FY 2004, it issued 31 SRB-related messages. In FY 2006, it issued ten SRB-related messages. Table 6.1 shows part of Milpers Message 07-141.[5] To read this table, an MOS 11B Infantryman in the rank of E-4 in Zone A qualified for an SRB multiplier of 1.0 and an 11B Infantryman in the same rank with SQIs of G and V qualified for a multiplier of 1.5. Notice that 11B E-4s in Zones B and C do not qualify for an SRB. It is useful to note that as of June 2007, personnel in Zone C in many MOSs were not eligible for an SRB. Furthermore, bonus amounts were capped, and the cap varied by MOS and zone.

Table 6.2 illustrates SRB amounts that an E-4 in Zone A and an E-5 in Zone B would receive based on the formula SRB = AOS × SRBM × MBP. The MBP is assumed to be $2,048 for the E-4 in Zone A and $2,571[6] for the E-5 in Zone B. The table indicates that a level 1 multiplier would provide an E-4 in Zone A an SRB of $6,144 for a three-year reenlistment. Theoretically, bonus amounts rise linearly with either AOS or SRBM such that a six-year reenlistment in an MOS with a level 6 multiplier would receive a bonus of $73,728. Bonus amounts are not this large, however. One reason is that no Army MOS has a multiplier of 6 (Table B.1) and few have multipliers above 3. Second, and more important, the Army places a ceiling on bonus payments (shown in the right-hand column of Table 6.1). In June 2007, most Army MOSs had a Zone A bonus cap of $10,000 (although some had caps as high as $30,000). Thus, individuals in level 1 multiplier skills would maximize their bonuses with a five-year reenlistment. Bonus caps discourage longer enlistments when they become binding. Although the bonus ceilings are generally higher in Zones B and C—around $15,000 to $20,000—these amounts are also quickly binding in those zones.[7]

Trends in Selective Reenlistment Bonuses, FY 2001–2007

The previous section makes clear the complicated nature of the Army's SRB program. This section summarizes the program using Army budget submission data as well as raw data provided by the Defense Manpower Data Center. The DMDC data are individual-level, longitudinal data on all Army entrants over the period FY 1988–2002 who were still on active duty at some time in the period FY 2001–2007. In addition to information from annual personnel records, from which reenlistment decisions may be inferred, the dataset contains information from

---

[5]   The full set of skills eligible for SRBs as of June 2007 is provided in Table B.1.

[6]   These are the FY 2008 monthly basic pay rates for an E-4 with over four years of service and an E-5 with over eight years of service, respectively.

[7]   It is important to note that these bonus caps are based on Army policy. Federal law specifies maximum SRB amounts. From FY 1991 to FY 2006, legislation limited the SRB to $60,000 (DoD, 2005, p. 627). Effective January 6, 2006, the SRB ceiling was increased to $90,000 (Volume 7A, Chapter 9 of DD 7000.14-R, September 2008).

Table 6.1
SRB Multipliers for Selected Skills, by MOS and Grade, June 2007

MOS	SQI	Multiplier							Bonus Cap (Zones)
		Grade							
		E-4	E-5	E-5	E-5	E-6	E-6	E-6	
		Zone							
		A	A	B	C	A	B	C	
All MOS	Special Forces (T)	0	4	4	4	4	4	4	$50K(A,B,C)
11B		1	1.5	2	0	1.5	2	2	$15K(A), $25K(B,C)
11B	Ranger (G), Airborne Ranger (V)	1.5	2	2.5	0	2	2.5	0	$15K(A), $25K(B,C)
11C		1.5	1.5	2	0	1.5	2	2	$15K(A), $25K(B,C)
13B	Parachute Qualified (P)	0	1	1	0	0	0	0	$10K(A), $15K(B,C)
13D		1.5	1	1	0	1	1	0	$10K(A), $20K(B,C)
13D	Parachute Qualified (P)	2.5	2.5	2.5	0	1.5	2	0	$10K(A), $20K(B,C)
13F		1.5	2	2	0	2	2	0	$10K(A), $20K(B,C)
18B		2.5	2.5	2.5	2.5	2.5	2.5	2.5	$30K(A,B,C)
18C		2.5	2.5	2.5	2.5	2.5	2.5	2.5	$30K(A,B,C)
18D		3.5	3.5	3.5	3.5	3.5	3.5	3.5	$30K(A,B,C)
18E		3.5	3.5	3.5	3.5	3.5	3.5	3.5	$30K(A,B,C)
18F		0	0	0	0	0	3.5	3.5	$30K(A,B,C)

SOURCE: Milpers Message Number 07-141, Army Human Resources Command, June 6, 2007.

NOTES: A, B, and C in this table refer to reenlistment Zones A, B, and C. The letters T, G, and P refer to Special Qualification Identifiers as defined in the text.

annual pay records about receipt of SRBs, SRB amounts, and SRB multipliers. Summary statistics about SRBs from this dataset are presented below.

Table 6.3 shows the number of SRB payments and SRB outlays reported in Army budget submissions over the FY 1998–2008 period.[8] Before FY 2005, the Army paid half of a new bonus award at the time of the reenlistment and the other half in annual installments spread evenly over the remaining period of the reenlistment. In FY 2005, the Army switched to lump-sum payments at the time of reenlistment. Table 6.3 therefore distinguishes between new bonus payments and installment payments. For the period since FY 2005, the column labeled

---

[8] With the exception of FY 2001 and FY 2002, all figures in Table 6.3 are actual figures for that fiscal year (rather than requests in budget submissions). But actual figures are unavailable for FY 2001 and FY 2002, so the figures reported for these years are requests taken from the Army's FY 2000 budget submission.

Table 6.2
SRB Amounts Based on SRB Formula

SRBM	Additional Obligated Service Years			
	3	4	5	6
E-4 in Zone A				
1	6,144	8,192	10,240	12,288
2	12,288	16,384	20,480	24,576
3	18,432	24,576	30,720	36,864
4	24,576	32,768	40,960	49,152
5	30,720	40,960	51,200	61,440
6	36,864	49,152	61,440	73,728
E-5 in Zone B				
1	7,713	10,284	12,855	15,426
2	15,426	20,568	25,710	30,852
3	23,139	30,852	38,565	46,278
4	30,852	41,136	51,420	61,704
5	38,565	51,420	64,275	77,130
6	46,278	61,704	77,130	92,556

NOTES: Amounts are in FY 2008 dollars. SRBM is the selective reenlistment bonus multiplier.

New Payment Rate shows the average bonus award amount in that year. For prior years, the average bonus amount equals twice the New Payment Rate. Average new bonus amounts ranged from about $9,000 in FY 1998 to $13,600 in FY 2008.

Table 6.3 indicates a downward trend in the number of reenlistees receiving SRBs over the FY 1999–2003 period followed by a trend increase thereafter. The number of new payments rose dramatically in FY 2005 and FY 2006. The total SRB budget increased from $103 million in FY 2003 to $708 million in FY 2006, a factor of almost 7.

No official DoD sources provide the percentage of reenlistees who receive SRBs. Such a percentage can be constructed by dividing the number of new SRB payments in Table 6.3 by the number of reenlistments, which is available from DoD. Table 6.4 shows the resulting percentages. Unfortunately, neither the number of reenlistments nor the number of new bonus payments is disaggregated by zone, so only the total percentage of reenlistments receiving SRBs can be constructed. Table 6.4 indicates a declining percentage of reenlistments receiving SRBs before FY 2004 following by a sharp rise in FY 2005–2006. The percentage estimated to be receiving SRBs declines in FY 2007–2008.

No official data source reports summary statistics on SRB multipliers, so the DMDC pay record data were used to do so. The advantage of pay record data is that summary statistics can be computed by reenlistment zone. Table 6.5 summarizes SRB multipliers in the period FY 2001–2004 and the period FY 2005–2007 for personnel who received an SRB. In the FY 2001–2004 period, at least half of the multipliers in each zone were either 0.5 or 1. Only a small percentage of the personnel in each zone received a multiplier of 3 or more. Bonus

Table 6.3
Army SRB Payments and Budgets, FY 1998–2008

Fiscal Year	No. of New Payments	New Payment Rate ($)	New Payment Outlay ($)	No. of Installment Payments	Installment Payment Rate ($)	Installment Payment Outlay ($ millions)	Total SRB Outlay ($ millions)
1998	6,277	4,530	28	17,295	1,284	22	51
1999	10,672	4,670	50	18,033	1,391	25	75
2000	15,288	4,775	73	22,825	1,419	32	105
2001	12,465	4,870	61	32,688	1,508	49	110
2002	7,946	4,513	36	45,379	1,189	54	90
2003	7,557	4,627	35	49,229	1,374	68	103
2004	18,117	4,799	87	44,126	1,269	56	143
2005	44,459	10,500	467	36,510	1,062	39	506
2006	65,156	10,600	691	16,720	1,062	18	708
2007	35,553	12,400	441	90,520	1,043	94	535
2008	47,123	13,600	641	43,032	1,043	45	686

SOURCE: Department of the Army, various years.

NOTES: FY 2001 and FY 2002 data are future-year estimates from the FY 2000 budget request. Data for other years are actual figures. FY 2006–2008 numbers exclude the Critical Skills Retention Bonus (CSRB).

Table 6.4
New Bonus Payments, Reenlistments, and Estimated Percentage
Receiving SRBs, FY 2000–2008

Fiscal Year	No. of New Payments	No. of Reenlistments	Percentage Receiving SRBs
2000	15,288	71,311	21.4
2001	12,465	64,982	19.2
2002	7,946	58,207	13.7
2003	7,557	54,151	14.0
2004	18,117	60,010	30.2
2005	44,459	69,512	64.0
2006	65,156	67,307	96.8
2007	35,553	69,777	51.0
2008	47,123	73,913	63.8

NOTE: Annual reenlistment counts for the Army were supplied by the Office of the Under Secretary of Defense for Personnel and Readiness (Officer and Enlisted Personnel Management).

Table 6.5
Percentage Distribution of Army SRB Multipliers, by Zone, FYs 2001–2004 and 2005–2007

SRBM	Zone A		Zone B		Zone C		All Zones	
	FY 2001–2004	FY 2005–2007	FY 2001–2004	FY 2005–2007	FY 2001–2004	FY 2005–2007	FY 2001–2004	FY 2005–2007
0.5	15	4	16	13	12	16	15	16
1	35	17	43	29	63	22	38	22
1.5	22	42	13	24	7	29	19	29
2	19	10	20	11	13	10	19	10
2.5	5	8	3	8	1	8	4	8
+3	3	19	5	15	3	15	4	15

SOURCE: Computed from raw data provided by Defense Manpower Data Center.

multipliers were increased significantly in the FY 2005–2007 period. Still, in this period, less than 20 percent of bonus recipients enjoyed a multiplier of 3 or more.

Reenlistment Bonus Program Overview, FY 2008

In June 2007, the Army introduced the Enhanced SRB (ESRB). This program represented a major departure from the previous SRB program in several ways. First, it eliminated SRB multipliers and use of a formula on which to base bonus amounts. Instead, bonuses were specified as fixed amounts that depended on MOS/SQI, rank, zone, and additional obligated service measured in one-year intervals up to five years. This meant that all individuals reenlisting for 37–48 months, for example, would receive the same SRB. Second, the Enhanced SRB program reduced the disparity in bonuses across MOSs. For example, notice from Table 6.1 that under the multiplier-based program, an 11B Infantryman in the rank of E-4 qualified for a multiplier of 1.0 whereas an MOS 13F Fire Support Specialist qualified for a 1.5 multiplier. Under the Enhanced SRB program, personnel in these two MOSs receive the same fixed dollar amount for the same additional obligated service. Third, the Enhanced SRB program tended to raise SRBs for personnel in Zones B and C relative to personnel in Zone A. Fourth, E-3 personnel in Zone A became eligible for an SRB. Under the older programs, E-3s were not eligible for reenlistment bonuses.

The Army continued the Deployed SRB program for personnel not otherwise qualified for an Enhanced SRB. The structure of the Enhanced SRB program is discussed below in more detail, followed by a discussion of the Deployed SRB program.

Enhanced SRB Program

The Army established three tiers of Enhanced SRBs. The first tier is based on a critical skills list; all MOSs on this list are eligible for the same SRB (which in turn depends on zone, rank, and AOS). A wrinkle here is that the Army actually established two bonus payment levels for personnel qualifying for a reenlistment bonus via the critical skills list. The wrinkle is that personnel who reenlist in the fiscal year of their contract expiration date—called the expiration of term of service (ETS) date—receive larger bonuses than personnel who reenlist before their contract expiration year.[9]

The second tier is based on a list of special critical skills. MOSs in this tier are eligible for larger SRBs than those in the first tier. For skills on this list, there is no difference in bonus payments based on time to ETS.

Finally, individuals who possess certain SQIs and are assigned to particular units are placed on a location critical skills list and may be eligible for an SRB even when personnel in their primary MOSs are not. Bonus amounts in this tier are larger than those in the first tier but smaller than those in the second tier.

The Army first implemented the Enhanced SRB program in June 2007 and made several changes to it between June and December. Using selected MOSs as examples, Table 6.6 illustrates the structure of the Enhanced SRB program as of December 2007.[10] To read Table 6.6, MOS 11B Infantryman was designated to be a critical skill (tier 1). Infantrymen who possess SQIs of G or V and were assigned to the 75th Ranger Regiment at Fort Benning were on the location critical skills list (tier 3) and qualified for a somewhat larger bonus than Infantrymen who did not possess these SQIs. Special Forces personnel (MOS 18B–MOS 18F) were deemed to have special critical skills (tier 2) and qualified for the largest Enhanced SRBs. Bonus amounts are discussed below.[11]

Also shown in Table 6.6 are the number of personnel in Zones A, B, and C in the MOSs eligible for the Enhanced SRB.[12] About 266,000 Army enlisted personnel were in Zones A, B, and C (corresponding to 2–14 years of service) in FY 2008. About 216,000, just over 80 percent of the total, were eligible for the ESRB based on criteria in effect in December 2007.

One feature of the Enhanced SRB program in effect in December 2007 was that all Zone A personnel in ranks E-3/E-6 and all E-4/E-7 personnel in Zones B and C were eligible for an SRB if their skill (MOS or SQI/unit) appeared in Table B.2. Before the Enhanced SRB program, neither E-3s nor E-7s were eligible for a reenlistment bonus. Furthermore, before the ESRB program, bonus eligibilities depended on rank as well as MOS and SQI/unit (Table B.1). Not conditioning eligibility on rank had the effect of significantly expanding the percentage of personnel eligible for an SRB.

The Enhanced SRB amounts first announced in June 2007 remained in effect until March 2008. Table B.3 lists Enhanced SRB amounts available during this period for MOSs on the critical skills list and the amounts available for MOSs on the special critical skills list.[13] The bonus amounts in Table B.3 represent a significant increase over amounts previously

---

[9]   Individuals may reenlist with as little as 17 months of service. However, bonuses are based on the amount of additional obligated service beyond the original contract expiration date. Just why the Army would want to pay larger SRBs to personnel who have largely completed their current enlistments is not entirely clear. One possible reason is that personnel who have largely fulfilled their current contracts have more experience, on average, than personnel who have not done so and are therefore more valuable to the Army. Early reenlistment is a signal of above-average taste for military service, and the Army may be taking advantage of this by "price-discriminating" against those with such higher taste.

[10]   See Milpers Message Number 07-344, December 12, 2007. The full set of skills eligible for an ESRB in December 2007 is listed in Table B.2.

[11]   Tables of ESRB amounts are provided in Appendix B.

[12]   Counts are based on personnel on active duty as of September 30, 2008, and are computed from individual-level data provided by the Defense Manpower Data Center.

[13]   The critical skills amounts shown are for personnel in their contract expiration year. To save space, critical skills amounts for those not at ETS are not shown, but these are somewhat smaller than those available for personnel at ETS. Also not

Table 6.6
Selected Skills Eligible for the ESRB and Overall Eligibility, December 2007

MOS	MOS Title	Number in Zones A, B, and C	Critical Skill	Special Critical Skill	Location Critical Skill	SQI, Unit
11B	Infantryman	37,414	Yes		Yes	G/V, 75TH RNGR
11C	Ind Fire Infantryman	4,246	Yes			
13B	Cannon Crewmember	6,168	Yes			
13D	Field Artillery Data System Specialist	1,977	Yes			
13F	Fire Support Specialist	4,456	Yes		Yes	G/V, 75TH RNGR
18B	Specialist for Weapons Sgt	907		Yes		
18C	Specialist for Eng Sgt	868		Yes		
18D	Specialist for Med Sgt	759		Yes		
18E	Specialist for Communications Sgt	922		Yes		
18F	Specialist for Intelligence Sgt	118		Yes		
Total number eligible for ESRB		216,274				
Total number not eligible		49,934				
Percentage eligible for ESRB		81.2				

SOURCE: Milpers Message Number 07-344, Army Human Resources Command, December 12, 2007.

available. For example, a Zone A 11B Infantryman in the rank of E-4 who reenlisted for four years would receive an SRB of about $8,000 under the June 2007 multiplier-based program (Table 6.2) but $14,500 for a 36–48 month reenlistment under the December 2007 Enhanced SRB program.

In March 2008, the Army reduced Enhanced SRB amounts across the board.[14] The revised bonus amounts are given in Table B.4. For example, the above 11B Infantryman who qualified for a $14,500 Zone A bonus for AOS of 36–48 months in the December 2007– February 2008 period now qualified for a bonus of only $9,500. In December 2007, an E-6 in Zone B on the critical skills list qualified for a bonus of $20,500 for 36–48 additional months of service; in March 2008, the same individual qualified for a bonus of only $15,500.

In addition to reducing bonus amounts, the March 2008 program revision restricted the ranks deemed to be in critical skills. One such rank restriction was that E-7s were eliminated from the lists. Furthermore, eligibilities were once again specified rank by rank for each MOS/ SQI. As an example, all E-3, E-4, E-5, and E-6 personnel in MOS 11B (Infantryman) who were in Zone A were designated to be in critical skills (and thus eligible for a bonus) but only E-6 personnel in MOS 13B (Cannon Crewmember) were so designated. The March 2008 program revision added some skills (MOSs and SQIs) to the three lists and eliminated others.

---

shown are amounts available for skills on the location critical skills list. These amounts may be found in Milpers Message Number 07-344, December 12, 2007.

[14] See Milpers Message Number 08-068, March 13, 2008.

These additions and deletions had less effect on overall bonus eligibility than the changes brought about by restrictions on rank.

We calculated the effect of the March 2008 eligibility revisions on the percentage of reenlistments qualifying for an Enhanced SRB. To do this, we first identified all personnel in our FY 2008 master personnel inventory in Zones A–C who had a change in obligated service of at least one year.[15] Approximately 30,000 personnel in Zone A had an AOS change of at least 12 months; the Zone B and Zone C counts were 15,500 and 7,800, respectively (Table 6.7). We then applied the December 2007 criteria for bonus eligibility, and the March 2008 criteria, to each individual. Results of these calculations are shown in Table 6.7. Consistent with Table 6.6, estimates are that about 80 percent of personnel with AOS of at least 12 months would have qualified for an ESRB had the December 2007 criteria for bonus eligibility remained in effect the whole fiscal year. Likewise, had the March 2008 criteria been in effect the whole year, a smaller percentage would have qualified for an ESRB (59 percent overall, 65 percent in Zone A, 53 percent in Zone B, and 48 percent in Zone C).

We calculated the percentage who were actually eligible for an Enhanced SRB during FY 2008 according to when the ETS changes occurred. We estimate that about 72 percent were eligible overall.

Deployed SRB Program, FY 2008

Soldiers not eligible for an Enhanced SRB may qualify for a Deployed SRB if they commit to additional service of at least six months beyond their ETS date while in a combat zone. An interesting feature of this bonus is that personnel may receive it for additional obligated service of at least six months rather than the one-year required of Enhanced SRB recipients. An additional wrinkle is that the ETS on which the Deployed SRB is based is the soldier's original contract expiration date and not a revised date pushed out by a stop-loss order.[16] Therefore, soldiers whose service has been involuntarily extended by a stop-loss order may receive the Deployed

Table 6.7
Eligibility for an Enhanced SRB, by Zone, FY 2008

	Number (%) Eligible Based on December 2007 Program Criteria	Number (%) Eligible Based on March 2008 Program Criteria	Number (%) Actually Eligible Based on Event Date	Number of AOS Changes of 12 Months or More in FY 2008
Zone A	24,398 (81.2)	19,606 (65.2)	22,618 (75.3)	30,054
Zone B	12,377 (80.0)	8,113 (52.5)	11,008 (71.2)	15,463
Zone C	6,295 (80.4)	3,741 (47.8)	4,839 (61.8)	7,827
All zones	43,070 (80.7)	31,460 (59.0)	38,465 (72.1)	53,344

---

[15] These numbers were constructed from raw individual-level data provided by DMDC. A reenlistment was deemed to have occurred if an individual's ETS at the end of FY 2008 was at least 12 months larger than the individual's ETS date at the end of FY 2007.

[16] A soldier is subject to a stop-loss order if his or her ETS date falls within the period (D − 90, R + 90), where D is the unit deployment date and R is the unit return date. When a stop-loss occurs, Army adjusts the individual's ETS date to R + 90.

SRB when they begin their involuntary stop-loss period if this period is at least six months. The Deployed SRB thus compensates additional service that is involuntary at the same rate as additional service that is voluntary.

Tables B.5 and B.6 provide the Deployed SRB amounts for the two time periods, December 2007 and March 2008. The Deployed SRB amounts are smaller than the Enhanced SRB program amounts. As it did to the Enhanced SRB amounts, the Army reduced the Deployed SRB amounts in March 2008.

Table 6.7 showed that roughly 80 percent of FY 2008 reenlistees were eligible for an Enhanced SRB under December 2007 criteria. This percentage dropped to 59 percent under March 2008 criteria. Overall, about 72 percent of FY 2008 reenlistees were eligible for an Enhanced SRB based on when they actually reenlisted. Adding in reenlistees who were eligible for a Deployed SRB but not an Enhanced SRB raises overall bonus eligibility to 92 percent under December 2007 bonus criteria, 82 percent under March 2008 criteria, and 89 percent based on actual reenlistment date during FY 2008.

Bonus Program Costs, FY 2008

Average bonus amounts and SRB program costs for FY 2008 may be estimated from the above information. Average Enhanced SRB amounts were calculated by assigning an Enhanced SRB amount to each individual who experienced an AOS change of at least 12 months in FY 2008, based on that individual's AOS and the bonus type for which the individual qualified. Results of these calculations are shown in the first panel of Table 6.8. The second panel of Table 6.8

Table 6.8
Average SRB Amounts, by Zone, FY 2008

	December 2007 Program	March 2008 Program	Actual Eligibility
Enhanced SRB			
Zone A	15,412	9,646	13,347
Zone B	21,278	14,644	19,261
Zone C	27,769	17,604	23,118
All zones	18,522	12,175	16,255
Deployed SRB			
Zone A	9,647	7,681	8,386
Zone B	11,065	8,723	9,600
Zone C	10,438	9,142	9,550
All zones	10,667	8,725	9,420
Any SRB			
Zone A	14,850	10,043	11,370
Zone B	19,925	13,291	15,295
Zone C	26,989	14,901	17,740
All zones	18,237	11,723	13,493

NOTE: Amounts are in FY 2008 dollars.

also provides average bonus amounts for individuals who were eligible for the Deployed SRB only (and had an AOS change of at least six months). Finally, the bottom panel shows the average bonus amount for all recipients.

If the December 2007 program had remained in effect the full year, the average bonus amount for Enhanced SRB recipients would have been $18,522, the average amount for Deployed SRB recipients would have been $10,667, and the overall average amount would have been $18,237. The actual average for FY 2008 is estimated to be $13,493, which is almost the same as the average amount for FY 2008 from Army budget data (Table 6.3).

Table 6.9 estimates the cost of the FY 2008 Army SRB program. The total estimated amount, based on aggregation of amounts for which individuals were eligible, was $639 million. This amount is very close to the Army's reported new payment outlay for FY 2008 of $641 million (Table 6.3).

Critical Skills Retention Bonus Program

One more bonus, thus far unmentioned, falls within the Army SRB program budget category. This bonus, introduced in FY 2006, is called the Critical Skills Retention Bonus (CSRB). It is largely targeted at senior enlisted personnel in selected skills who are approaching retirement eligibility at the 20-years-of-service mark.[17] The CSRB was originally aimed at Special Forces personnel (to include personnel in any MOS with an SQI of T) but more recently has been expanded to include personnel in ten other MOSs. This program is smaller in cost and in numbers of personnel than the main SRB program. According to Army budget documentation, 406 personnel received this bonus in FY 2007 at a cost of $31 million; budget plans call for 526 recipients in FY 2009 at a cost of about $39 million.[18]

Table 6.10 shows the CSRB amounts available for Special Forces personnel.[19] The interesting feature here is the payment structure. Two additional years of commitment beyond the

Table 6.9
Army SRB Program Cost, by Zone, FY 2008

	December 2007 Program	March 2008 Program	Actual
Zone A	417	260	310
Zone B	283	162	209
Zone C	196	92	118
All zones	900	515	639

NOTE: Amounts are in millions of dollars.

---

[17] In most skills eligible for the CSRB, a requirement for eligibility is that the individual have between 19 and 23 years of service. In three skills, MOS 35M (Human Intelligence Collector), MOS 35P (Cryptologic Linguist), and MOS 89D (Ordnance Disposal), personnel with less than 19 years of service may qualify for the CSRB, depending on rank.

[18] Department of the Army, various years.

[19] Lesser amounts are available to other, non–Special Forces personnel. See Army Milpers memorandum 08-324, December 19, 2008, for details.

Table 6.10
Critical Skills Retention Bonus Amounts, by Length of Commitment (Special Forces)

Two Years	Three Years	Four Years	Five Years	Six Years
18,000	30,000	50,000	75,000	150,000

NOTES: Amounts as of December 2008. See Milpers Message Number 08-324. Special Forces MOSs include 18B, MOS 18C, MOS 18D, 18E, and MOS 18X. The bonus is also available to personnel in any MOS with an SQI of T. Amounts are in FY 2008 dollars.

individual's current obligation qualifies the individual for an $18,000 bonus and a three-year additional commitment qualifies for a $30,000 bonus. The marginal bonus value of the third year is thus $12,000. The marginal bonus values for commitments beyond three years are $20,000, $25,000, and $75,000, respectively. In fact, the bonus for a six-year commitment is twice as large as the bonus for a five-year additional commitment. This convex bonus structure provides very strong incentives for longer commitments.

Methodology and Data for the Army Reenlistment Model

Chapter Six described the Army's Selective Reenlistment Bonus program and documented the substantial expansion in this program over the FY 2001–2008 period. The question naturally arises whether this program had an effect on Army retention. The primary purpose of this chapter is to provide an empirical assessment of this question. To do so, it studies Army reenlistment during the period FY 2002–2006 and estimates the reenlistment effects of Army SRB multipliers. Reenlistment effects include effects on the likelihood of reenlistment and effects on the length of reenlistment. The analysis also addresses the reenlistment effect of deployments in support of Operation Iraqi Freedom and Operation Enduring Freedom as well as the effects of Army stop-loss policies. The cost of additional person-years induced by SRBs is calculated and compared with other policies for increasing person-years. SRBs are found to be more cost-effective than other policies for expanding person-years. Chapter Eight presents the estimated effects of reenlistment bonuses on reenlistment in the other services as well as the Army, building on a recent study of reenlistment, deployment, and bonuses (Hosek and Martorell, 2009).

Data

The data for the analysis presented below were provided for the most part by the Defense Manpower Data Center. The primary dataset consists of a longitudinal file containing all Army non-prior service entrants over the period FY 1988–2002. In addition to information from each entrant's contract and accession records, the dataset contains annual information on each individual from DMDC's enlisted master record files.[1] Individuals are tracked annually from the entry year through FY 2007. About one million non-prior service personnel entered the Army over the period FY 1988–2002.

DMDC supplemented these data with detailed individual-level data on deployments, stop-loss status, SRB multipliers and amounts, and certain other information. In particular, DMDC provided monthly information indicating whether each individual in the master dataset was under a stop-loss order and deployment start and end dates for each deployment the individual experienced during the FY 2001–2008 period.

Because of the massive size of the combined datasets, it was not computationally feasible to undertake an analysis of Army reenlistment that used data on all personnel. Therefore, 24 Army Military Occupation Specialties were selected for analysis (Table 7.1). MOSs were

---

[1] Data are as of September 20 of each year.

Table 7.1
Army MOS, Number of Entrants, and Percentage of 24-MOS Sample

MOS	MOS Title	Number of Entrants FY 1988–2002	Percentage
11B	Infantryman	121,751	25.23
11C	Indirect Fire Infantryman	14,530	3.01
13B	Cannon Crewmember	33,063	6.85
13F	Fire Support Specialist	11,086	2.3
14S	Avenger Crewmember	5,721	1.19
14T	Patriot Missile Crewmember	5,683	1.18
15T	UH–60 Helicopter Repairer	5,270	1.09
15U	CH–47 Helicopter Repairer	2,605	0.54
19D	Cavalry Scout	18,910	3.92
19K	M1 Armor Crew	29,381	6.09
21B	Combat Engineer	22,539	4.67
25Q	Multi–Channel Transmission Systems Operator/Maintainer	11,860	2.46
25U	Signal Support Systems Specialist	13,229	2.74
31B	Military Police Officer	31,110	6.45
35F	Intelligence Analyst	663	0.14
35P	Cryptologic Linguist	7,205	1.49
63B	Light–Wheel Vehicle Mechanic	25,725	5.33
63H	Track Vehicle Repairer	5,359	1.11
63M	Bradley Vehicle Systems Mechanic	6,397	1.33
88M	Motor Transport Operator	23,237	4.82
92A	Automated Logistical Specialist	20,619	4.27
92F	Petroleum Supply Specialist	19,426	4.03
92G	Food Service Specialist	24,673	5.11
92Y	Unit Supply Specialist	22,500	4.66
Total		482,542	100

NOTE: The counts in this table refer to number of entrants into each MOS during the FY 1988–2002 period, not the number remaining in service by FY 2001, which will be smaller because of losses occurring before FY 2001.

selected partly on the basis of size and partly to provide representation across the spectrum of Army MOSs. About one-quarter of the individuals in these 24 occupations are in MOS 11B (Infantryman). These 24 occupations constitute about half of all personnel in the master database.

Although the master file contains data for FY 2007, the reenlistment analysis uses data from the FY 2002–2006 period. The reenlistment models estimated below use lagged deployment data, so FY 2001 is lost as a result of lagging. FY 2007 data are not used, for two reasons. First, those data were provided soon after the end of FY 2007, and some of the dates needed to fully identify all FY 2007 reenlistments do not appear to have been updated. FY 2007 reenlistments are therefore undercounted. Second, the Army introduced the Enhanced SRB program in June 2007; this program so radically changed the way it does reenlistment bonuses that evaluation of its effect is best left for a future analysis.

Retention

As Chapter Six explained, federal legislation has established three zones of eligibility for reenlistment bonuses. Zone A includes the period from 17 months to 72 months of service (YOS 2–6 where YOS denotes year of service) and corresponds to the end of the first term. Zone B includes the period from 72 months to 120 months of service (YOS 7–10) and corresponds to the end of the second term, and the period from 120 months to 156 months of service (YOS 11–14) is Zone C and corresponds to the end of the third term.[2] Although personnel may reenlist as early as at 17 months of service (independent of the length of their initial enlistment), reenlistment bonuses are computed on the basis of the additional obligated service beyond the current contract service date known as the Expiration of Time in Service date. Before June 2007, federal law and Army policy required that reenlistees commit for at least three additional years to qualify for an SRB.

Although an individual who enlists in the Army for the modal contract length of four years could conceivably reenlist 31 months before his or her contract expiration date and receive a reenlistment bonus, most reenlistments occur within a year of the ETS date. For that reason, the analysis focuses on reenlistments that occur within one year of ETS and ignores early reenlistments.[3] Table 7.2 reports the number of personnel reaching ETS in our 24-MOS sample during the period FY 2001–2006. (Reaching ETS means that the individual had less than 12 months to the ETS date at the start of the fiscal year.)

Although a change in obligated service of at least three years is required for receipt of an SRB, the Army defines AOS of at least two years to be a reenlistment and AOS of less than two years to be a contract extension. This analysis maintains that convention. Table 7.3 reports the ETS reenlistment, extension, and total retention rates of the personnel in the 24-MOS sample during the FY 2002–2006 period.

Retention in all three zones jumped between FY 2001 and FY 2002. Retention in each zone then declined relative to the zone's FY 2002 retention. Still, by the end of the time period, retention in Zones B and C still exceeded FY 2001 retention. There is a caveat to interpretation of the Zone A trend in Table 7.3. The decline in Zone A retention in FY 2005–2006 is partly an artifact of the dataset from which the rates reported in Table 7.3 were constructed. Since

---

[2]   For personnel who are in the last day of their 72nd month of service, the zone into which they fall is based on the hour of the reenlistment!

[3]   According to our data, early reenlistments—reenlistments before the 12-month window preceding the ETS date—make up about 11 percent of total reenlistments. But the early reenlistment rate is higher among deployed personnel (15 percent) than nondeployed personnel (9 percent). The higher rate of early reenlistment among deployed personnel no doubt reflects the fact that bonuses received while in a combat zone are not subject to income tax.

Table 7.2
Number Reaching ETS in the 24-MOS Sample,
by Zone, FY 2001–2006

Fiscal Year	Zone A	Zone B	Zone C
2001	22,690	8,097	3,702
2002	26,115	10,491	4,382
2003	17,458	7,507	2,914
2004	30,075	12,104	3,810
2005	27,999	11,274	3,597
2006	15,161	9,085	2,783

NOTE: Reaching ETS means that the individual had less
than 12 months to the ETS date at the start of the fiscal
year.

Table 7.3
Reenlistment, Extension, and Total Retention Rates in the 24-MOS Sample Among the FY 1988–2002
Entry Cohorts, FY 2001–2006

Fiscal Year	Zone A			Zone B			Zone C		
	Reenlistment	Extension	Total	Reenlistment	Extension	Total	Reenlistment	Extension	Total
2001	35.1	7.1	42.2	45.1	14.7	59.8	68.4	6.5	74.9
2002	46.3	8.7	55.0	52.6	16.2	68.8	73.0	9.7	82.7
2003	42.2	10.2	52.3	46.1	17.8	63.9	71.9	9.3	81.2
2004	39.4	9.1	48.5	50.3	12.6	62.9	68.4	9.9	78.3
2005	34.6	7.4	42.0	54.8	8.4	63.2	71.3	6.1	77.4
2006	33.7	6.6	40.3	52.8	7.2	60.0	68.6	6.3	74.9

NOTE: The base for these rates is the at-ETS counts in Table 7.2.

the 24-MOS sample contains only individuals who signed contracts to enter the Army before
FY 2003, FY 2004–2006 do not contain reenlistments of individuals who signed contracts
after FY 2002, entered the Army, and subsequently reenlisted. That is to say, there is some
undercounting of Zone A reenlistments in FY 2004–2006. Since most Army enlistments are
for three or more years, the undercounting is more likely to have reduced the Zone A FY 2006
retention rate reported in Table 7.3 than either the FY 2004 or FY 2005 rates. Since no post-
FY 2002 entrants would have reached Zones B or C by FY 2006, there is no undercounting of
reenlistments in these zones.

Stop-Loss and Deployment

In FY 2002, the Army began to implement stop-loss policies designed to prevent soldiers
whose contracts were expiring from separating from the Army. Soldiers are subject to stop-loss
if they have an ETS date that falls within a time interval that begins 90 days before the start
of a unit deployment (referred to as D − 90) and ends 90 days after the return of the unit to
its home base (referred to as R + 90). When a stop-loss order goes into effect, the soldier's ETS

date is adjusted to R + 90 plus one day, and soldiers are not eligible to separate voluntarily before this date.

Table 7.4 shows the percentage of all personnel (including those not at ETS) in the 24-MOS sample that had a stop-loss flag at some time during the fiscal year and the percentage that had a deployment of any duration during the year, over the period FY 2001–2006. There were no stop-losses in FY 2001 and few in FY 2002. The percentage of all personnel experiencing a stop-loss then rose to 7.2 percent in FY 2003, 18.4 percent in FY 2004, and 20.5 percent in FY 2005 before declining to 14 percent in FY 2006. Personnel in Zone C were less subject to stop-loss than personnel in Zones A and B. In those zones, personnel in Zone A experienced slightly more stop-loss than personnel in Zone B.

Only a handful of personnel were deployed in FY 2001 or FY 2002. Deployments rose dramatically in FY 2003 and thereafter, reflecting the mobilizations for OIF/OEF. In FY 2004, over half of the personnel in our 24-MOS sample had a deployment. Deployment rates of personnel in Zone A were about 5 percentage points higher than rates for individuals in Zone B, and 10 percent higher than rates for individuals in Zone C.

Table 7.5 shows the frequency distribution of cumulative deployment time for all personnel in the 24-MOS sample. There were no deployments in FY 2001 and few in FY 2002. By FY 2003, about 43 percent of the personnel had experienced some deployment time; by FY 2005,

Table 7.4
Percentage of All Personnel in the 24-MOS Sample with Stop-Loss or Deployment, by Zone, FY 2001–2006

Fiscal Year	All Zones		Zone A		Zone B		Zone C	
	Stop-Loss	Deployment	Stop-Loss	Deployment	Stop-Loss	Deployment	Stop-Loss	Deployment
2001	0	0.6	0	0.6	0	0.6	0	0.7
2002	0.6	6.7	0.5	6.5	1.1	7.3	1.3	7.1
2003	7.2	40.4	7.1	41.9	9.7	37.2	4.6	34.8
2004	18.4	51.5	21.2	55.2	17.2	46.0	8.5	42.3
2005	20.5	43.5	24.9	46.7	19.9	42.6	11.2	36.8
2006	14.0	48.0	17.8	53.3	15.6	47.4	8.5	41.8

NOTE: The base for these rates is defined as all personnel in the 24-MOS sample who were still in service as of a given fiscal year and in a given zone.

Table 7.5
Percentage Distribution of Cumulative Months of Deployment in the 24-MOS Sample, FY 2001–2006

Months	2001	2002	2003	2004	2005	2006
0	100	93.6	58.3	34.9	24.2	18.4
1–12	0	6.7	40.6	56.9	50.2	40.4
13–24	0	0	1.1	8.2	25.1	37.8
25–36	0	0	0	0.1	0.4	3.4

NOTE: The base for these rates is defined as all personnel in the 24-MOS sample who were still in service as of a given fiscal year and in a given zone.

almost 66 percent had experienced some deployment time, and by FY 2006, over 80 percent had. The percentage with deployment time exceeding one year rose from 1 percent in FY 2003 to about 41 percent in FY 2006.

Trends in Selective Reenlistment Bonus Multipliers

Chapter Six discussed the evolution and structure of the Army SRB program. During the period under study, the Army based the SRB on the formula SRB = SRBM × LOR × MBP. Multipliers range from 0 to 6; during the observation period, the Army adjusted them periodically in half-multiplier increments. Notice that each half-unit change in the multiplier represents a half-month's basic pay per year of reenlistment.

During the study period, the Army varied multipliers by MOS, rank, zone, and assignment location. Furthermore, multipliers in many MOSs varied by subcomponents of skill called Special Qualification Identifiers. In FY 2003, the Army also introduced an SRB for deployed personnel. The Deployed SRB kicks in if it exceeds the normal SRB for which the individual is qualified. Thus, an individual who is deployed, and not otherwise qualified for an SRB, would receive the Deployed SRB.

Figure 7.1 shows the average SRB multiplier by month computed from the DMDC pay record data. The averages are based on the individuals who reenlisted in the given month. A multiplier value of 0 is imputed to an individual if the pay record data do not indicate that the individual received an SRB. (The numbers are therefore "unconditional" SRB multiplier averages.) Figure 7.1 indicates that the Army reduced multipliers between FY 2002 and mid–FY 2003, held them low until the end of FY 2004, and then began increasing them sharply

Figure 7.1
Average SRB Multiplier, by Zone

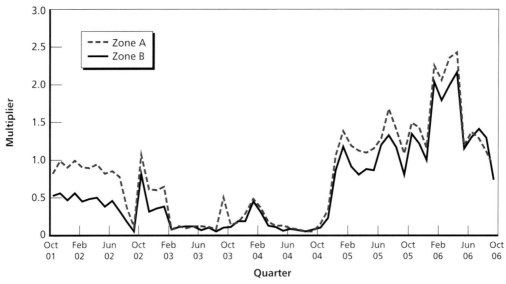

SOURCE: Computed from DMDC pay record data.
RAND *MG950-7.1*

in FY 2005. Notice the sharp increase in multipliers in both zones in the first three months of calendar 2006. The Army raised multipliers for all reenlistees in both zones during this period. It then rescinded the increases. According to DMDC pay record data, the average Army multiplier for all of FY 2006 was 1.53 in Zone A and 1.29 in Zone B.

As Chapter Six discussed, whenever the Army changes its SRB program, the Army Human Resources Command announces the change in a memorandum. To see how consistent the DMDC pay record data were with the Army's SRB program announcements, the "generic" SRB multiplier available in each announcement for each MOS-rank-zone combination in the 24-MOS sample was recorded, and the multiplier was attached to each reenlistee in the 24-MOS sample.[4] The average generic multiplier was computed by zone.

Figure 7.2 compares the average generic Zone A multiplier from Army SRB announcements with the averages constructed from the DMDC data. To the extent that the averages constructed from DMDC data account for the fact that some personnel might qualify for bigger multipliers than the generic multiplier, the DMDC averages should exceed the generic averages, and Figure 7.2 indicates that they do. For the most part, the two series move together. But Figure 7.2 points to a problem with the DMDC data. In the second half of FY 2004, the DMDC averages fall below the generic averages. After consultation with DMDC and Army officials, it was revealed that there were electronic difficulties in reporting the data from the Army to the Defense Financial Accounting Service (DFAS) and from DFAS to DMDC. As

Figure 7.2
Average Zone A SRB Multiplier Based on DMDC Pay Record Data Compared with Average Zone A Multiplier from Army Bonus Memoranda

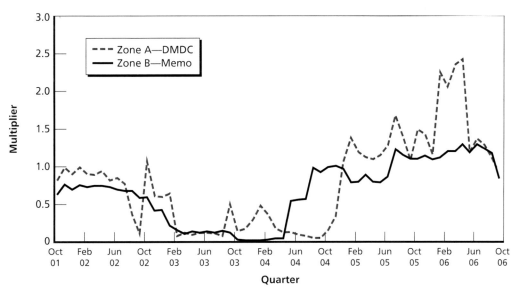

---

[4] "Generic" means the basic multiplier available to all personnel in the MOS; it ignores larger multipliers that might be available for possession of certain SQIs or assignment to certain locations. It also ignores the Deployed SRB.

a result, not all bonus payments and multipliers were recorded in this six-month period. Of course, there may be underreporting elsewhere, but this seems to be the most prevalent period.[5]

The next section discusses formulation of an SRB variable for use in the reenlistment analysis and considers how to deal with the underreporting issue.

Reenlistment Models and Estimation Results

This section presents the results of econometric analysis of individual reenlistment decisions of the personnel who reached ETS during the FY 2002–2006 period. The goal here is to obtain estimates of the effects of SRB multipliers, deployment, and stop-loss on reenlistment. One way to obtain estimates is to develop a "structural" model of reenlistment decisions and then infer the effects of pay and other changes via simulation. Asch et al. (2008) provide an example of this approach. This approach does have two costs, however. One is the relative complexity of the model and estimation procedure. The other is the difficulty in accommodating many control variables in the analysis. Because the research goal is to obtain direct estimates of the effects of SRB multipliers, deployment, and stop-loss policies on reenlistment, we eschew a structural approach in favor of straightforward "reduced-form" estimation of the reenlistment effects of these variables using standard econometric techniques.[6]

Specification of Key Variables

The SRB Multiplier. The key variable of interest in this analysis is the SRB multiplier for which an individual was eligible at the time of the reenlistment decision. It is not easy to decide which SRBM should be assigned to each individual. SRBMs are observed in the DMDC pay records for those who reenlist and receive an SRB but not for those who reenlist and do not receive an SRB or those who do not reenlist.[7] One possibility is to assign the latter two groups bonus multipliers based on multipliers provided in various Army Milpers messages. This approach was not feasible, for two reasons: (1) the complexity of the Army SRB program in this period and (2) the lack of electronic documentation of the program. Rather, it was necessary to adopt an empirically based assignment strategy. Namely, DMDC pay record data were used to compute the average SRBM and the median SRBM by cell in our data, where the cell is defined in two different ways, for the reasons discussed in more detail below.[8] The first

---

[5] Figure 7.2 shows a significant increase in the SRB average based on DMDC data during the first four months of calendar 2006 but no such increase in the SRB average based on Army memos. In fact, Milpers Message Number 06-007 (January 5, 2006) provided an immediate SRB multiplier increase of 0.5 for all personnel in any zone with an ETS date between January 6, 2006, and September 30, 2006, and an increase of 1.5 for personnel with an ETS date in FY 2007. This message emphatically stated that these were temporary increases and would expire on April 30, 2006. The DMDC average SRB series in Figure 7.2 reflects these temporary adjustments. The memo multiplier series is flat during this period, because the Army did not adjust multipliers announced in other messages to incorporate these temporary changes.

[6] Since the dependent variable is a binary indicator for reenlistment, models are estimated by the probit method, a form of regression for binary outcome data. Readers who are unfamiliar with probit analysis are referred to Cameron and Trivedi (2005) for a treatment of probit analysis.

[7] The DMDC pay records provide SRBMs only for reenlistees who actually received SRBs. Reenlistees who did not receive an SRB were assigned an SRBM of zero.

[8] As noted above, SRB amounts changed frequently and varied according to duty location and detailed (within-MOS) skill. For this reason, measuring the effect of an SRB within a simple choice (stay/leave) framework is problematic. For example, an individual in MOS 11B may be eligible for an SRB if he or she agrees to serve in a particular location and unit

approach defines a cell by MOS-zone-rank-deployment-time period combination (cell) in our data. The second approach defines a cell by MOS-zone-rank-time period combination (cell) but independently of deployment. In this assignment process, reenlistees who, according to DMDC pay record data, did not receive an SRB were assigned an SRB multiplier and an SRB amount of zero. Each individual was then assigned the average or median value of SRBM in his or her particular cell.[9]

As shown by the discussion above that accompanies Figure 7.2, this assignment process will understate SRB multipliers available to personnel in the second half of FY 2004, but it should accurately measure the multipliers available in other periods. Measurement error in an independent variable is, of course, problematic, and it is well known that when a variable is measured with error, and the measurement error is random, regression analysis tends to understate the true effect of the variable (that is, estimates are biased downward in absolute value).[10] But, since the measurement error here is not random but is concentrated in FY 2004, it will be compensated for (at least in part) by the inclusion of an FY 2004 time effect in the model. A check on this claim is to estimate the model excluding data for FY 2004. In fact, when this is done, the estimates are not materially different from those shown below. Therefore, measurement error does not appear to be a major concern.

Regarding the definition of SRBM, the values for MOS, zone, and rank were those at the time of the reenlistment decision, whereas deployment was based on whether the individual had any deployment in the current fiscal year. In some cases, an individual in the "deployment" cell had already returned from deployment, perhaps as long as 11 months earlier, but in others, the individual was still deployed.

As mentioned, one version of our bonus variable defines the cell by MOS, rank, time period (month), and term (in effect, zone) but not by deployment. The other version defines the cell by the same variables and deployment, where deployment is deployment in the year of the reenlistment decision. Chapter Eight also uses these two alternative approaches to define the bonus variable in the analysis of reenlistment for all four services.[11] We use two definitions of the cells by which we compute the average SRBM, because each approach introduces different biases to the estimated effect of bonuses on reenlistment. We therefore use both approaches as a way to bracket the true estimate (i.e., come up with a lower and upper bound).

---

but is otherwise ineligible. If the full choice set available to each soldier were observable, one could estimate a richer model that makes use of the variation in the incentives (i.e., bonuses or bonus multiples) associated with the different choices. Since we do not observe the full choice set available to each soldier, we have resorted to the simpler procedure of assigning to each soldier at a point in time the average bonus multiplier received by observationally equivalent individuals who reenlisted and received a bonus at that time.

[9] To be clear, averages and medians were based on those who reenlisted; nonreenlistees were assigned values based on cell definitions.

[10] For a discussion of measurement error in regression analysis, see Cameron and Trivedi (2005, pp. 899–922).

[11] The definitions used in this chapter differ from those used in Chapter Eight because of differences in the empirical methodology. In Chapter Eight, the version of the bonus variable that depends on deployment depends on deployment in the month of the reenlistment decision, whereas in this chapter, it is deployment in the year of the reenlistment decision. Making the deployment window to be the same month as the reenlistment decision (as in Chapter Eight) means that the bonus computed for the cell will correspond to the bonus available to the deployed service member, whereas having a longer window increases the chance that the computation of the bonus for the deployment cell includes both currently deployed and returned-from-deployment individuals.

When we define the bonus variable only in terms of MOS, rank, and term but not deployment, we introduce measurement error in the bonus variable, especially for the Army, because bonuses differ by deployment status, as discussed in Chapter Six. That is, a deployment bonus may be paid to individuals who reenlist when deployed and are not eligible for an SRB. The amount is averaged into the bonus for the specialty, whereas it is payable only to deployed members of the specialty. The introduction of this error can be expected to bias the estimate of the bonus effect on reenlistment toward zero.

On the other hand, when we allow the definition of the bonus variable to also depend on deployment status, we introduce a problem that, we believe, leads to an upward bias in the estimate of the bonus effect. To see this, note that bonuses are higher for personnel who reenlist while deployed if they can receive a deployment bonus (although they are not eligible for a SRB) and because of the combat zone tax exclusion (CZTE). CZTE increases the value of a SRB. For instance, at a 25 percent marginal tax rate, a $100 bonus is worth $75 if taxed and $100 if excluded from tax. Further, in the Army data discussed in Chapter Eight, nearly everyone who makes a decision when deployed decides to reenlist. Of soldiers making a reenlist/leave decision when deployed to a hostile area in the decision month, 93 percent reenlist. This extraordinarily high percentage may be due to stop-loss policies that in effect prevent a "leave" decision but allow a "stay" decision. For personnel who want to leave, the decision to exit is postponed until stop-loss is lifted 90 days after deployment. Also, service members may strategically time their reenlistment to occur during deployment to take advantage of the higher bonuses and preferential tax treatment. Thus, defining SRBMs conditional on deployment can produce a large upward bias in the bonus effect. We return to this in the discussion of the results in this chapter and in Chapter Eight.

Summarizing, this chapter, like Chapter Eight, constructs the bonus variable for cells defined by MOS-zone-rank-deployment-time period and for cells defined by MOS-zone-rank-time period. In this chapter, deployment is defined as deployment at some time within the past 12 months including the current month, whereas in Chapter Eight, deployment is defined as deployment in the month of the reenlistment decision. The approach used in this chapter (focusing on annual data) in effect averages the bonus paid at reenlistment to members who reenlist when deployed and the bonus paid at reenlistment to individuals who were deployed in the past year but are not currently deployed. The approach in Chapter Eight is likely to result in bonus effect estimates that bracket those in this chapter—and that is what proves to be the case. In our view, both definitions as well as approaches have limitations, but taken together they should enable us to triangulate on a credible range of the true bonus effect. As the discussion at the end of Chapter Eight shows, the results from Chapters Seven and Eight turn out to be consistent and fairly close to one another.

Variables for Stop-Loss and Deployment. The other key variables relate to deployment and stop-loss. The reenlistment effects of deployment and stop-loss were allowed to interact with one another. The following combinations of deployment and stop-loss status during the reenlistment decision year were allowed: (1) deployed without a stop-loss, (2) deployed with a stop-loss initiated during the ETS year (referred to below as "first stop-loss"), (3) deployed with a stop-loss that was not initiated during the ETS year (i.e., a stop-loss that was continued from the previous year), and (4) not deployed but with a stop-loss. Individuals in the last category consist of personnel who have returned from a deployment during which a stop-loss order was in effect and are almost all personnel for whom a stop-loss order had been in effect in the previous year. The reference category for the effects of these variables is personnel who were neither deployed nor had a stop-loss during the reenlistment decision year.

In addition to current deployment status, cumulative past deployment time is also allowed to affect reenlistment. To measure the effects of cumulative past deployment time, three indicators of past deployment are included in the estimation: (1) cumulative deployment of less than 13 months, (2) 13–24 months of past deployment time, and (3) more than 24 months of past deployment time.

MOS Effects, Time Effects, and Other Control Variables. A problem plaguing estimation of the effect of SRBs on reenlistment is bias arising from correlation between SRBs and unobservable factors that cause variation in reenlistment. One source of correlation is persistent differences in retention by MOS, time, and other factors. For example, If bonus managers have a target retention rate for every MOS but persistent differences in retention exist across MOSs because of working conditions, exposure to danger, and other factors, bonus managers will set higher bonuses the more the actual retention rate falls below the target rate. A retention analysis that ignores persistent differences in retention by skill will understate the effect of SRBs.[12] As another example, if all bonuses rise (fall) during periods of generalized retention shortfalls (overages), bonus effects will likewise tend to be biased downward.

To reduce the extent of omitted variables bias, the estimated models therefore include a host of variables that are likely to be related to retention. Since the analysis uses data on 24 MOSs, the estimated models include 23 MOS dummies. The excluded (reference MOS) is MOS 11B. The analysis includes controls for the rank an individual holds at the time of reenlistment. The excluded (reference) rank is E-4. In principle, SRB multipliers are set by zone, so the analysis should include zone dummies. But even within reenlistment zones, there is significant variation in retention by year of service. Therefore, the models include controls for years of service rather than zone.

When Army personnel deploy, they deploy in units from their permanent bases. One might naturally expect reenlistment to vary by unit because of differences in operating tempo, in exposure to danger in theater, in leadership and morale, and in other unobservable factors. Although each individual's exact Army unit was not observable, the permanent base to which the individual was assigned was. The models therefore controlled for permanent base of assignment with the following caveat. Most bases had more than 10,000 personnel, but a number of bases had very few personnel. To keep the number of bases manageable, small bases in the same geographic area were grouped together (so, for example, the smaller bases in Georgia were grouped together, as were the bases in Alaska). The estimated models included 21 dummies for military bases in the United States. The excluded (reference) group is assignment outside the United States.

The models also control for demographic characteristics, including age, race and ethnicity, gender, dependents, and quality category. Three controls are included for race-ethnicity: black, Hispanic, and other race. The excluded (reference) group is white. Five indicators are included for an individual's number of dependents in the year of reenlistment decision: one, two, three, four, and five or more. For about 10 percent of the data, the number of dependents is unknown. The models therefore include an indicator for known dependents. The excluded (reference) category is personnel with unknown dependents.

---

[12] Hosek and Martorell (2009) estimate reenlistment models for all four services with and without MOS controls; they find that estimates without MOS controls are much smaller and in some cases even negative.

The models also include a high-quality indicator. High-quality personnel are defined as those who have a high school diploma and score 50 or above on the Armed Forces Qualification Test.

The models do not include controls for military pay (or military pay relative to civilian pay) and unemployment. Military pay rose by about 10 percent in real terms over the data period (Simon and Warner, 2007) and civilian unemployment trended downward over the period. Higher pay should lead to higher retention and lower unemployment should reduce it. But because the time period is short (the analysis uses FY 2002–2006 data), and because these factors are highly correlated with time, their inclusion will not improve the estimated models nor give plausible estimates of their effects. Therefore, they are excluded from the analysis.[13]

Even with inclusion of such an extensive set of controls to reduce the extent of bias arising from omitted variables, biased estimation may not be completely avoided. Hosek and Martorell (2009) and Goldberg (2001) analyze the bonus-setting process and show conditions under which "endogeneity" bias will still be present after accounting for persistent factors. Endogeneity bias arises when remaining random shocks to retention within an MOS and time period also affect bonus multipliers. The common view is that the bias is negative as a result of the feedback from adverse (positive) retention shocks and bonus increases (decreases) to offset (neutralize) those shocks.[14] If such reverse causality exists, standard regression (or probit) methods will tend to understate the true effects of bonuses.

Instrumental variables methods exist for dealing with endogeneity. Instrumental variables are correlated with SRB multipliers but not retention. Estimates obtained with IV procedures have been shown to be consistent (unbiased in large samples). The IV approach to the problem of SRB endogeneity has been limited by the fact that it is difficult to find variables that are plausibly correlated with SRB multipliers but not retention.[15] In the end, we have to hope that there is enough exogenous variation in SRB multipliers in the data to yield plausible positive estimates of SRB effects.[16]

Annual Data Models and Monthly Data Models

The above discussion of key variables implicitly assumed that the empirical analysis would be based on one observation per individual in the individual's ETS year. This is referred to as

---

[13] Simon, Negrusa, and Warner (forthcoming) study first-term retention in all four services over the FY 1991–2003 period and include controls for relative military pay and unemployment. Both factors are related to retention in the expected direction: Higher values of either pay or unemployment serve to increase retention.

[14] Hosek and Martorell (2009) develop several models of bonus-setting within an MOS and time period. Since bonus managers typically set bonuses by skill at the start of a time period and before retention is observed, one wonders how bonuses can be endogenous. The answer is that there may be autocorrelation between random factors causing a retention shock in the last period and random factors causing a retention shock in the current period. Autocorrelated shocks therefore lead to correlation between bonuses set at the start of a period and subsequent bonus levels. But they also provide several examples of situations in which bonus effects will be biased upward, not downward. Interested readers are referred to Appendix A of their study.

[15] Interested readers are referred to Simon and Warner (2009) for an analysis of the Air Force enlistment bonus program over the period FY 1998–2001. This study used two instruments for enlistment bonuses—training cost and training days—to obtain consistent estimates of the effects of bonuses on the length of the initial Air Force enlistment.

[16] Hosek and Martorell (2009) show that estimated SRB effects are consistent when (1) random retention shocks are not autocorrelated and (2) there are random shocks to target retention by MOS and time period. The latter shocks are shocks to demand. Demand shocks that cause SRB multiplier variation help identify the effects of SRB multipliers on retention.

an annual data analysis. The next subsection presents empirical estimates of the effects of key variables based on this approach.

There is an alternative approach. Individuals may reenlist at any time during the ETS window, and reenlistments begin to occur in significant numbers around the 15-month mark before ETS. Therefore, it is possible to estimate models of reenlistment based on monthly data. This approach conceptualizes the reenlistment decision during the ETS interval as a sequence of monthly decisions. In each month in the event window, an individual decides whether to reenlist after comparing the value of reenlisting immediately with the "option" value of deferring the decision to another month (which in turn depends on both the value of a future reenlistment and the value of a future separation). These values depend on many factors, including the SRB multiplier available today and the SRB multiplier the individual expects to prevail in the future. When the SRB multiplier is low today but is expected to be higher in the future, the value of reenlisting now is low, and the option value of delaying the reenlistment is high. Therefore, the likelihood of reenlistment in the current month will be low. Conversely, the reenlistment rate will be high in the current month if the current multiplier is also high but is expected to be lower in the future.

Since the key policy factors of interest—SRB multipliers, deployment status, and stop-loss status—are all observed monthly, the monthly data approach arguably makes better use of the data than do the annual data models. The cost of this approach is that it expands the size of an already large dataset by a factor of 12. As a consequence, it was not possible to estimate monthly data models using the full dataset. As a compromise, four occupations were selected for analysis: MOS 11B (which accounts for 25 percent of the data), MOS 13B, MOS 13F, and MOS 21B (see Table 7.1 for a description of these MOSs.) These MOSs were selected because they are large and because they are combat-related, with working conditions and exposure to risk that are likely to be similar to the environment faced by 11B personnel. Again, these models are estimated by the probit method. Also, in the monthly model (although not in the annual model), we show results only for the case where the average SRBM is computed by MOS-zone-rank-deployment-time period combination (cell). That is, we do not show the case where the cell is not conditional on deployment status.

Treatment of Extensions

Table 7.3 indicates that over the FY 2001–2006 period, there was about one contract extension for every four to five reenlistments. Furthermore, the extension rate was higher in FY 2003 than in other years, indicating some sensitivity of extensions to time-related factors. A long-standing issue in reenlistment analysis is how to treat contract extensions.[17] To the extent that contract extensions are free choices and are made in response to compensation policies in effect at the time of the decision, it is natural to think of individuals choosing among three alternatives—reenlist, extend, or separate—at the time of the retention decision and to model the chosen outcome as the one with maximum value in a three-choice framework. Indeed, it is likely that extensions will be high relative to reenlistments when SRBs are low, because in this circumstance reenlistment has a lower value than a contract extension. Furthermore, extensions may increase if personnel expect higher bonuses in the future. Goldberg and Warner

----

[17] Goldberg (2001) provides a lengthy discussion of this issue.

(1982) studied Navy retention using a three-choice approach and found that higher SRB multipliers raise Navy reenlistments while reducing both extensions and separations.

Use of a three-choice approach to Navy retention in the late 1970s made sense, because Navy extensions during this period were largely voluntary decisions. But the three-choice approach is less compelling here because many of the Army extensions during the FY 2002– 2006 period were not based on voluntary choices on the part of soldiers but resulted from involuntary adjustments to ETS dates made by the Army in the wake of unit deployments and stop-loss orders. In this circumstance, it makes more sense to treat extensions as deferred reenlistment decisions. Individuals who experience an ETS date change of less than two years are not considered to have yet made a reenlistment decision. The binary decision that is studied, therefore, is the ultimate decision to reenlist or separate.

The following two subsections provide and discuss estimated reenlistment effects of the key variables in the analysis. The first subsection presents results from the annual data models; the second subsection contains estimates from monthly data. All models are estimated by maximum likelihood probit; each coefficient in the tables below indicates how the probability of reenlistment changes as a result of a change in the variable in question (i.e., the variable's marginal effect). Full estimation results are provided in Tables C.1 and C.2, respectively.[18]

Results of Annual Data Analysis

Recall from Table 7.3 that reenlistment rates trended downward from FY 2002–2006. The estimated models included annual time effects to capture this trend. Figure 7.3 plots the time effects estimated from the Zone A and Zone B models, respectively. These trends thus hold constant all of the other factors in the estimated models. Figure 7.3 indicates that, ceteris paribus, the probability of reenlistment fell by over 20 percentage points between FY 2002 and FY 2006—a precipitous decline. In other words, this is the retention decline that would have been expected in the wake of an improving economy and OIF/OEF in the absence of any adjustments to bonuses and other policies that might improve retention. The observed retention decline (Table 7.3) was not nearly this large, suggesting that policy adjustments did have an effect.

One such policy effect, as is known from the bonus program history review in Chapter Six, was adjustments to SRB multipliers, which increased over the time period and particularly in FY 2005–2006. Above, we discussed econometric issues relating to construction of the SRB multipliers from pay records. In particular, it was shown that conditioning the SRB multiplier on deployment status may lead to upward-biased estimates but that not conditioning it on deployment status may lead to downward-biased estimates. Because of the uncertainty about the best way to measure SRB multipliers, estimates based on both methodologies are shown in Table 7.6.

---

[18] Because individuals' retention decisions may be related to one another, the observations are not fully independent of one another. Treating them as if they were independent leads to understatement of standard errors of estimates and overstatement of the statistical significance of the estimates. "Clustering" of standard errors helps guard against this bias. Clustering requires choosing factors that are related to the dependence in retention decisions. Because individuals are assigned to military bases and interact so much with other personnel assigned to that base, the factor most related to the dependence in retention decisions is likely to be the base to which an individual is assigned. Also, retention decisions may all be affected by common shocks at a particular time. We therefore cluster our standard errors on base and fiscal year. For more on clustering, see Cameron and Trivedi (2005, pp. 829–851).

Figure 7.3
Time Effects in Reenlistment, Relative to FY 2002

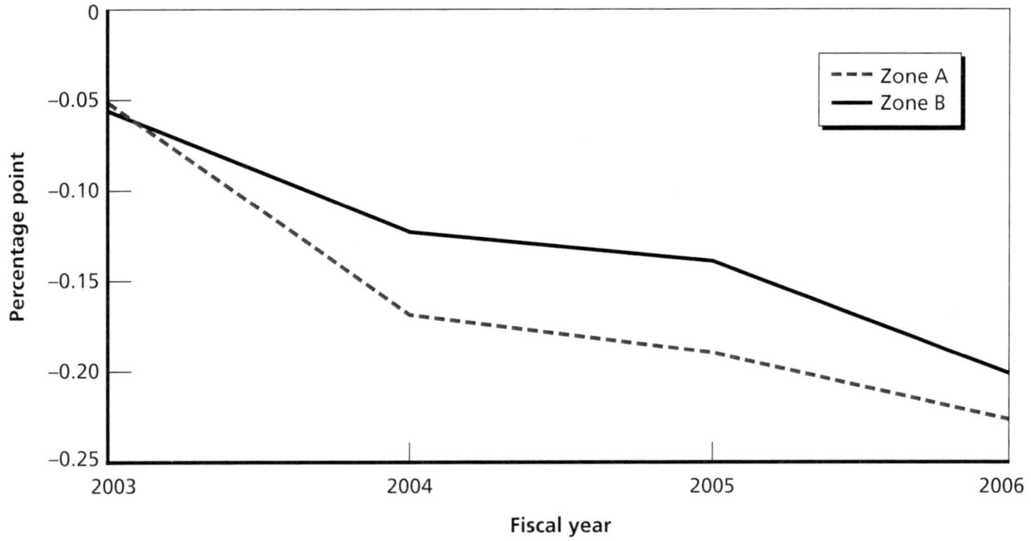

RAND *MG950-7.3*

According to the various estimates in Table 7.6, larger values of the SRB multiplier are associated with higher rates of reenlistment. And, as suspected from the above discussion, conditioning the multiplier on deployment status yields larger estimates of bonus effects. In the model that combines data from Zones A and B and conditions on deployment status, a one-multiple increase is associated with a 5.6 percentage point increase in the probability of reenlistment. In the combined model that does not condition on deployment status, a one-multiple increase is associated with a 3.5 percentage point increase in the reenlistment rate.

Estimating the models separately for Zones A and B gives somewhat larger estimates of SRBM effects, 5.9 percentage points and 7 percentage points, respectively, when the multiplier is not conditioned on deployment status, and 3.9 percentage points and 4.4 percentage points, respectively, when it is. The Army average SRB multiplier rose by a multiple of 1 between FY 2002 and FY 2006. Our most conservative estimate suggests an average reenlistment rate increase of 3.5 percentage points resulting from this increase; the most optimistic estimate suggests that Zone B reenlistments could have risen as much as 7 percentage points. The estimates obtained here are remarkably similar to those found in two recent studies sponsored by the Army Research Institute.[19] Further discussion and implications of these estimates for the cost-effectiveness of SRBs is provided below in the section titled the "The Cost-Effectiveness of Selective Reenlistment Bonuses."

---

[19] The first study, by Hogan et al. (2005), used Army data over the period FY 1990–2000. Estimates from a pooled model based on data for all Army MOSs indicate that a one-multiple SRB increase raises Zone A reenlistment by 6.4 percentage points and Zone B reenlistment by 4.4 percentage points. The second study, by Moore et al. (2006), used Army data over the period FY 2001–2004. The pooled model estimates from this study indicate that a one-multiple SRB increase raises Zone A reenlistment by 3.6 percentage points and Zone B reenlistment by 3.9 percentage points. Both studies also provided estimates disaggregated by MOS. These estimates show a great deal of variability—some much higher than the (within-study) pooled model estimates and some much lower.

Table 7.6
Estimated Marginal Effects of Key Variables, Annual Data Model, by Zone

| | SRBM Varies by Deployment Status? | | | | | |
| | Zones A and B | | Zone A | | Zone B | |
	Yes	No	Yes	No	Yes	No
SRB multiplier	0.056*	0.035*	0.059*	0.039*	0.070*	0.044*
	(0.012)	(0.012)	(0.014)	(0.014)	(0.010)	(0.013)
Deployed, no stop-loss	0.277*	0.288*	0.270*	0.283*	0.277*	0.283*
	(0.018)	(0.017)	(0.020)	(0.019)	(0.010)	(0.013)
Deployed, first stop-loss	−0.034	−0.015	−0.053	−0.032	0.019	0.036
	(0.041)	(0.041)	(0.042)	(0.043)	(0.040)	(0.041)
Deployed, continued stop-loss	0.193*	0.201*	0.181*	0.190*	0.210*	0.216*
	(0.049)	(0.048)	(0.055)	(0.055)	(0.03)	(0.032)
Not deployed, stop-loss	−0.220*	−0.218*	−0.204*	−0.202*	−0.240*	−0.238*
	(0.016)	(0.016)	(0.018)	(0.017)	(0.020)	(0.021)
<13 months cumulative deployment	−0.105*	−0.106*	−0.103*	−0.104*	−0.098*	−0.097*
	(0.023)	(0.023)	(0.024)	(0.024)	(0.02)	(0.022)
13–24 months cumulative deployment	−0.170*	−0.171	−0.164*	−0.165*	−0.174*	−0.171^
	(0.029)	(0.029)	(0.030)	(0.030)	(0.03)	(0.032)
>24 months cumulative deployment	−0.164*	−0.162*	−0.115	−0.113	−0.279*	−0.275*
	(0.056)	(−0.057)	(0.077)	(0.077)	(0.08)	(0.078)
No. of observations	119,956	119,956	91,468	91,468	28,488	28,488
Mean reenlistment rate	0.489	0.489	0.456	0.456	0.596	0.596

NOTE: Standard errors are clustered by fiscal year and military base.
* Denotes statistical significance at the 5 percent level.

Consider now the estimated effects of deployment and stop-loss. Estimates in Table 7.6 indicate that deployed soldiers reenlist at a higher rate than nondeployed soldiers. One interesting contrast is between deployed soldiers without a stop-loss and nondeployed soldiers without a stop-loss. Overall, and in both zones separately, the estimated differential between these groups is approximately 0.28–0.29, or 28–29 percentage points.[20]

Another interesting contrast is between deployed soldiers with a stop-loss in their ETS year (but before their original ETS date) and nondeployed soldiers with a stop-loss (the latter group consisting of personnel who have returned from a deployment and are free to leave at

---

[20] Notice in Table 7.6 that the various deployment status effects are slightly larger in models that condition the SRB multiplier on deployment status than in models that do not. This was to be expected, because a conditional multiplier is positively correlated with deployment. (The average multiplier of deployed personnel is about 0.2 higher than the average multiplier for nondeployed personnel.) As a result, part of the variation in reenlistment that the conditional models attribute to the SRB multiplier is attributed to deployment status in models that do not condition the multiplier on deployment.

R + 90 but are still under a stop-loss order). For Zones A and B together, the estimate reenlistment difference between these groups is 18.6 percentage points (−0.034 − (−0.22) = 0.186) when conditioning the SRB multiplier on deployment status and 20.3 percentage points (−0.015 − (−0.213)) when not doing so. Again, holding constant stop-loss status, deployed personnel reenlist at a significantly higher rate than the nondeployed.

Holding constant deployment status, soldiers subject to stop-loss are much less likely to reenlist than soldiers who are not subject to stop-loss. However, this association does not imply causality. Rather, soldiers are subject to stop-loss when their initial ETS falls within the dates of a deployment and their ETS dates are reset to a time after the deployment (i.e., an extension has occurred). Stop-loss is turned on precisely because the soldier did not reenlist before the original (contractual) ETS date. It is useful to note that although a stop-loss flag is an indicator of a reduced likelihood of reenlistment, not all soldiers subject to stop-loss leave after returning from a deployment.

Estimates in Table 7.6 suggest that the more cumulative deployment time a soldier has had, the lower the likelihood of reenlistment. The combined model estimates indicate that soldiers who have had between one and two years of past deployment time, or more than two years of past deployment time, are about 17 percentage points less likely to reenlist than soldiers with no past deployment time. The effect of more than two years of past deployment time is particularly large for Zone B personnel (−28 percentage points).

Analysis of Monthly Data

Turning now to the monthly model analysis, Figure 7.4 shows the monthly "hazard" rate for reenlistment of Zone A 11B personnel, by months to ETS.[21] The monthly hazard of reenlistment of 11B personnel in Zone A at ETS over the five-year period FY 2002–2006 averaged 0.027, or 2.7 percent, meaning that conditional on not reenlisting before that time, there was a 2.7 percent probability that an individual reenlisted in a given month. The cumulative hazard rate—that is, the reenlistment rate in Zone A—averaged 0.36, or 36 percent. The rates for Zone B were 0.055 and 0.60 and for Zone C, 0.087 and 0.79.

Notice that the hazard rises significantly 12 months before ETS, declines between ETS-12 and ETS-1, and jumps at ETS. The time path of the hazard reflects a combination of heterogeneity and uncertainty in the reenlistment decision. Individuals who have a high taste for military service, and therefore a high value of current reenlistment, reenlist early on, whereas individuals who are less certain about their futures, and who therefore place a high option value on delaying the decision, prefer to wait (procrastinate?) until the last minute before making their reenlistment decision.

An interesting observation from Figure 7.4 is that the downward trend in reenlistment across all 24 MOSs (Table 7.3) is not readily apparent here for MOS 11B personnel. Since the ramp-up in Army bonuses in the FY 2005–2006 period was concentrated in the Combat Arms MOSs, bonus policy could explain part of this. Furthermore, such MOSs may attract personnel who do not mind arduous and risky duty as much as other personnel might. Consistent with this casual inspection of the series in Figure 7.2, estimated time effects in the monthly models (found in Table A.2) are generally small and are not statistically significant.

---

[21] The *hazard rate* is the probability that a reenlistment occurs in a month, given that one has not occurred before that month.

Figure 7.4
MOS 11B Zone A Reenlistment Hazard, by Months to ETS

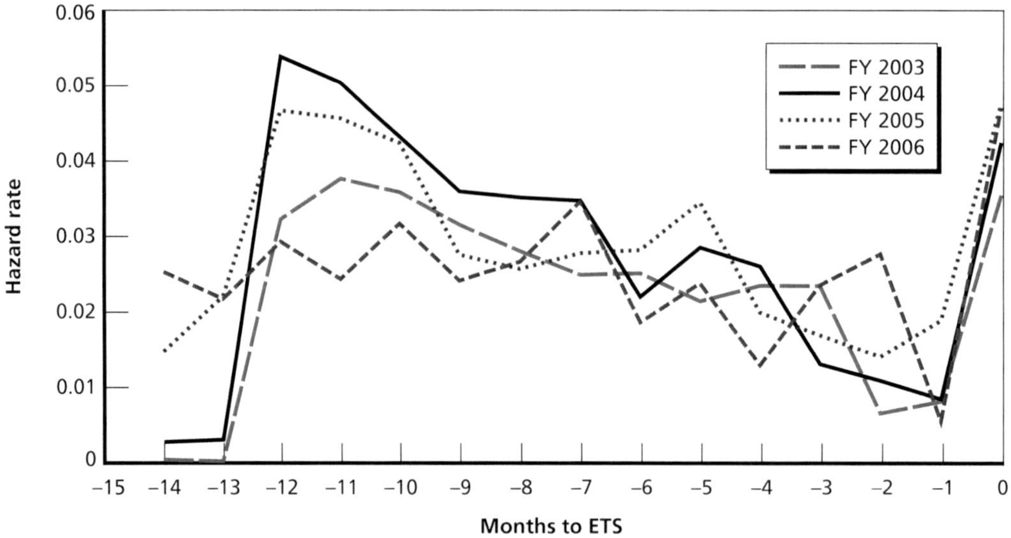

Estimated marginal effects of key variables on the monthly reenlistment hazard are provided in Table 7.7.[22] Consistent with estimates from annual data, these estimates indicate a significant positive effect of the SRB multiplier. Compared with the nondeployed who are not under a stop-loss, the deployed who are not under a stop-loss are more likely to reenlist each month in the ETS window. However, only the Zone B estimate is statistically significant. But, consistent with the annual data estimates, personnel who are not deployed but are under a stop-loss are significantly less likely to reenlist than the nondeployed who are not under a stop-loss. The same is true of those who are deployed but are under a stop-loss. More past deployment time is also estimated to exert a negative influence on reenlistment.

The estimates in Table 7.7 show how the *monthly* hazard *for reenlistment* (as defined above) is affected by each variable in question. However, it is more informative to calculate the effect of each variable on the total (cumulative) reenlistment rate over the full reenlistment event window. This cumulative effect is the marginal effect on reenlistment in the 15th month before ETS plus the marginal effect in the 14th month before ETS weighted by the probability of not reenlisting in month 15 plus the marginal effect in the 13th month before ETS weighted by the probability of not having reenlisted before that month, and so forth.[23] Estimates of cumulative

----

[22] As stated above, the monthly models were estimated by probit. Although use of a probit model for time-to-event data is unconventional, Simon, Negrusa, and Warner (2009) have found in previous analysis of GI Bill use with a panel of veterans that both logit and probit estimation of the probability of an event at time *t* with controls (time dummies) for baseline hazards gives virtually the same estimated marginal effects of regressors (and their significance levels) as estimation of an exponential survival time regression with the same controls. This method, of course, ignores the panel aspect of the data.

[23] The cumulative effect can be calculated from the recursion formula

$$R_t = \sum_{t=-12}^{0} [m \times r_t \times (1 - R_{t-1})]$$

Table 7.7
Estimated Monthly Reenlistment Effects of Key Variables, by Zone

	Zones A and B		Zone A		Zone B	
	Marginal Effect	Standard Error	Marginal Effect	Standard Error	Marginal Effect	Standard Error
SRBM	0.0050*	0.0011	0.0047*	0.0011	0.0046*	0.0019
Deployed, no stop-loss	0.0017	0.0014	0.0012	0.0013	0.0077*	0.0029
Not deployed, stop-loss	−0.0143*	0.0010	−0.0137*	0.0010	−0.0186*	0.0020
Deployed, stop-loss	−0.0123*	0.0012	−0.0107*	0.0011	−0.0132*	0.0025
<13 months cumulative deployment	−0.0016	0.0015	−0.0017	0.0014	−0.0017	0.0024
13–24 months cumulative deployment	−0.0011	0.0016	−0.0008	0.0015	−0.0020	0.0027
>24 months cumulative deployment	−0.0034	0.0042	−0.0006	0.0068	−0.0110*	0.0044
Sample size and mean monthly reenlistment hazard						
No. of observations	558,312		425,178		129,849	
Mean probability	0.031		0.026		0.047	

NOTES: The definition of SRBM is conditional on deployment status. Standard errors are clustered by base and year.

* Denotes statistical significance at the 5 percent level.

reenlistment effects based on this procedure are shown in Table 7.8 for the Zone A and Zone B coefficients in Table 7.7. (Combined model calculations are suppressed to save space but are virtually the same as the Zone A estimates provided here.) Also shown are estimates from models that use FY 2005–2006 data only. These models were estimated to determine whether

Table 7.8
Reenlistment Effects of Key Variables over the Full ETS Window, by Zone

	FY 2002–2006 Data		FY 2005–2006 Data	
	Zone A	Zone B	Zone A	Zone B
SRB multiplier	0.050*	0.034*	0.042*	0.029*
Deployed, no stop-loss	0.016	0.054*	0.058*	0.037*
Not deployed, stop-loss	−0.170*	−0.163*	−0.203*	−0.138*
Deployed, stop-loss	−0.127*	−0.111*	−0.154*	−0.106*
<13 months cumulative deployment	−0.009	−0.014	0.018	0.013
13–24 months cumulative deployment	−0.008	−0.015	0.02	0.014
>24 months cumulative deployment	−0.007	−0.111*	−0.048	−0.032

* Denotes statistical significance at the 5 percent level.

where m is the (monthly) marginal effect of a given variable, r_t is the base reenlistment hazard in the tth month before ETS, and R_t is the cumulative reenlistment rate up to period t in the ETS window. (By definition, $R_{-13} = 0$ in this calculation.)

the effects of deployment and stop-loss are more pronounced when estimated with data from this restricted period only.[24]

Larger SRBs are associated with a higher probability of reenlistment. The estimated (cumulative) effect of a one-multiple increase in the SRB multiplier ranges from 0.029 to 0.050. These estimates are somewhat smaller than those found in Table 7.6, which are based on annual data and where we define the SRBM conditional on deployment status and, in fact, they are very much in line with estimates from past studies.[25]

The combined effects of deployment and stop-loss are directionally similar to the estimates found in Table 7.6, which are based on annual data. However, they indicate generally smaller effects of deployment and stop-loss than those estimated previously. Unlike the annual data results, cumulative past deployment does not appear to influence reenlistments here.

Other Notable Effects

Both sets of models controlled for demographics (race, age, gender, and number of dependents), quality (high-quality if the AFQT score ≥50 and the candidate has a high school diploma or better; low-quality if the AFQT <50 or the educational status is less than a high school diploma), military rank, years of military service, military occupation specialty, and permanent military base of assignment. Interested readers are referred to Tables C.1 and C.2 for exact estimates of the effects of these variables. Here, we briefly summarize their estimated effects, which, in most cases, are very consistent between the annual and monthly models.

First, rank differentials in the probability of reenlistment are large. In the annual model for Zone A, for example, E-5 personnel have over a 30 percentage point higher probability of reenlisting than E-4 personnel. These large rank effects no doubt capture differences in current and expected future compensation by rank as well as other nonmonetary rank-related differences. Furthermore, personnel who achieve a higher rank by a given year of service are clearly better job matches for the military. Although the current analysis does not incorporate data before FY 2002, differences in reenlistment by rank may well have increased because of the post-2000 restructuring of basic pay, which widened the differentials by rank.

Second, holding rank constant, the probability of reenlistment declines with years of service. This result reflects the fact that, conditioning on rank, personnel with longer service have more limited career prospects (lower current compensation and more limited expected future promotion opportunities) than personnel of the same rank who have not served as long.

Third, there is substantial variation in the probability of reenlistment by MOS and by permanent base of assignment or assignment location.[26] With some exceptions, the probability of reenlistment is higher in other MOSs than in the reference MOS in our regressions, MOS 11B (see, for example, Table C.1). Given the arduous nature of this MOS, such an outcome was

---

[24]   Since there was no stop-loss and little deployment before FY 2003, their effects may be masked in an analysis that uses these years. Limiting the analysis to FY 2005–2006 data permits the effects of these variables to reveal themselves over time.

[25]   See Warner and Asch (1995), Goldberg (2001), and Asch, Hosek, and Warner (2007) for reviews of past studies of the reenlistment effects of SRBs. Past studies provide estimates of the effect of a one-multiple change in the SRB multiplier in the range of 2–6 percentage points, with a median estimate of about 3 percentage points. The recent study by Moore et al. (2006) was not included in these surveys; as noted above, that study provides estimates that are close to those provided here.

[26]   There were too many bases in our data to include all base dummies, so we combined the smaller bases that were in the same geographic areas. Thus, "US GA" in the appendix tables combines all installations in Georgia other than Fort Benning and Fort Stewart.

not unexpected.[27] There is significant variation also by permanent base or location of assignment (the reference group of which is assignment to a non-CONUS base in Europe or Asia).

Fourth, demographic characteristics matter for reenlistment, with number of dependents being the most influential. In the annual data model for Zone A reenlistment, for example, personnel with one dependent are estimated to be about 15 percentage points more likely to reenlist than personnel with no dependents, and personnel with five or more dependents are 41 percentage points more likely to reenlist than personnel with no dependents. Differences in reenlistment by dependents status dwarf estimated effects of other demographic attributes. Males, for example, are only about 5 percentage points more likely to reenlist than females, and high-quality personnel are about 4 percentage points less likely to reenlist than low-quality personnel.

Selective Reenlistment Bonuses and the Length of Reenlistment

The analysis thus far has looked at the effect of SRB multipliers on the probability of reenlistment. But just as enlistment bonuses can affect the length of enlistment (Chapter Four), reenlistment bonuses may affect the length of reenlistment.[28] This subsection uses the data on reenlistees in the 24-MOS sample to estimate how SRBs affect LOR.[29]

Army personnel may reenlist for a period of two to six years and, before June 2007, a reenlistee had to reenlist for at least three years to qualify for an SRB. This policy requirement, by itself, ought to lead to longer reenlistments among bonus-eligible reenlistees than among those who are not eligible for a bonus. Since each one-multiple increase in the SRB multiplier gives the reenlistee an extra month's basic pay per year of reenlistment, an SRB multiplier increase also ought to have a positive incentive effect on LOR. However, if the rate at which reenlistees are willing to trade longer reenlistments for larger bonuses diminishes as the multiplier increases, these increases will have smaller LOR effects as they increase. That is, we should expect to see LOR respond more positively to the SRB multiplier when it is low to begin with than when it is high.

A key policy factor is likely to diminish the responsiveness of LOR to the multiplier as the multiplier increases—ceilings on SRB amounts. Before FY 2005, most Army reenlistees confronted a bonus ceiling of $20,000. But beginning in December 2004, the Army began to separate bonus ceilings into four categories that depended MOS, SQI, and unit.[30] Personnel in the first category, which was based on MOS only, had their bonus ceilings reduced, Zone A from $20,000 to $10,000 and Zone B from $20,000 to $15,000. The $20,000 ceiling was continued for personnel in the second category, which was based on MOS and SQI and included Rangers and paratroopers. The third category consisted primarily of Special Forces personnel;

---

[27] It is useful to note the importance of controlling for MOS effects in models of reenlistment. Bonuses tend to vary inversely with an occupation's reenlistment rate; omitting occupation dummies from the estimation causes estimated bonus effects to be biased downward. In fact, estimated bonus effects are often negative when MOS effects are not controlled for.

[28] This chapter uses LOR to denote the period of a reenlistment decision. It is the same as, and is used interchangeably with, additional obligated service.

[29] Two recent analyses of the length of reenlistment are provided by Tsui et al. (2005) and Hogan and Simonson (2007). As this section proceeds, estimates provided below will be compared with estimates from these studies.

[30] See Milpers Message Number 04-355, December 30, 2004.

this category enjoyed a ceiling increase to $30,000. Finally, the fourth category consisted of personnel in any MOS or unit with an SQI of "T," for which the ceiling increased to $40,000. Despite ceiling increases for some personnel, the majority faced lower ceilings after December 2004.

The presence of a bonus ceiling, when it is binding, is expected to have a negative effect on the length of reenlistment. To see why, consider the following examples, where C denotes the bonus ceiling, MBP denotes monthly basic pay, and SRBM denotes the SRB multiplier. Since the bonus amount is given by the formula SRBM × LOR × MBP, the LOR that maximizes the bonus is given by LOR* = C/(MBP × SRBM). If the ceiling is $20,000, SRBM is 1.0, and MBP is $1,800 (about the average MBP of an E-4 with four years of service during the FY 2002–2006 period), LOR* equals 11.2 years. But since reenlistments cannot exceed six years for the purpose of bonus computation, the $20,000 ceiling is nonbinding. If the multiplier is 2.0, the bonus ceiling is not binding until an LOR of 5.6 years. If the multiplier increases to 3, the maximum bonus is achieved for an LOR of 3.7 years (or 44 months).

Likewise, a reduction in the ceiling from $20,000 to $10,000 means that an individual who has an SRBM of 1.5 would receive a $10,000 bonus for any LOR exceeding 44 months. Clearly, from this discussion, the presence of a bonus ceiling can have a perverse effect on LOR if an SRB multiplier increase shortens the LOR required to receive the maximum SRB. Hogan and Simonson (2007) analyze recent Navy data and find evidence that bonus ceilings reduce LOR in that service.

Evidence that SRB multipliers and bonus ceilings do in fact influence Army LOR is provided in Table 7.9. This table contains two panels. The upper panel shows the average months of reenlistment of the personnel in our 24-MOS sample who reenlisted, by zone, in the full FY 2002–2006 period. The lower panel shows LOR averages for FY 2002–2004 only, the period during which most personnel faced a bonus ceiling of $20,000.

First, it is apparent from Table 7.9 that up to a multiplier of 1.5, multiplier increases are associated with longer LOR. Over either time period, the average months of reenlistment for those who did not receive an SRB was 36 for Zone A and 38 for Zone B. Zone A personnel who had an SRB multiplier of 0.5 reenlisted for 40 months on average. Zone B personnel were apparently more responsive to this multiplier increase, as they had an average LOR of 47. Among Zone A personnel, a one-half multiple increase from 0.5 to 1 increased the average LOR by about three months, as did an increase from 1 to 1.5.

Table 7.9
Average Months of Reenlistment, by Zone and SRB Multiplier

Zone	SRB Multiplier							
	0	0.5	1	1.5	2	2.5	3	3.5
FY 2002–2006								
A	36	40	43	46	46	44	42	35
B	38	47	53	51	49	45	42	38
FY 2002–2004								
A	36	40	43	45	49	51	47	47
B	38	47	51	52	54	46	47	49[a]

[a] This average is based on only 23 observations. All others are based on at least 100 observations.

Second, Table 7.9 provides evidence that bonus ceilings matter. With a bonus cap of $20,000, bonus ceilings become binding when the multiplier exceeds 2. In fact, LOR averages based on the FY 2002–2004 period, when $20,000 was the ceiling for most personnel, begin to decline after a multiplier of 2.5 for Zone A personnel and 2 for Zone B personnel, and multiplier increases are associated with lower LOR beyond these points. Notice that beyond a multiplier of 2, the average LOR declines when FY 2005–2006 reenlistments are included in the averages. Such a decline would be expected because many, if not most, personnel making reenlistment decisions in the FY 2005–2006 period confronted lower ceilings (although some did enjoy higher ceilings).

The averages reported in Table 7.9 do not, of course, hold constant other influences on the length of reenlistment. To obtain estimates that do, regressions were estimated between months of reenlistment and dummies for the various values of SRB multiplier and the other numerous controls employed above. Before examining the results, two methodological issues need to be discussed. The first relates to the range of the dependent variable. Because LOR cannot take on values below 24 or above 72, it is censored from below and from above. Ordinary least squares estimates are potentially biased in the presence of censoring. But consistent estimates are easily obtained with the Tobit method.[31]

The second problem relates to the potential for sample selection bias arising from correlation between unobservable factors that affect the probability of reenlistment and unobservable factors affecting the length of reenlistment. Since personnel who, in fact, reenlist are "selected" from the population of personnel eligible to reenlist, estimates of the bonus effects on LOR, for example, will be biased if they materially change the composition of the group that reenlists. A bonus increase that causes many more reenlistments among personnel who have a lower (unobservable) taste for service could lead to understated (overstated) estimates of bonus effects on LOR if those with a lower taste are predisposed to shorter (longer) reenlistments.

Ways to deal with sample selection bias in two-step processes were first developed by economist James Heckman; his methods are discussed by Cameron and Trivedi (2005, pp. 546–555). Heckman's method involves estimating a probit equation for the probability of reenlistment (as done above in this chapter), creating a variable called the Inverse Mills Ratio (IMR) from this first-step estimation, and including IMR in the second-stage equation for LOR. A significant coefficient on the IMR variable indicates a nonzero correlation between factors affecting the probability of reenlistment and factors affecting length of reenlistment.

Models were estimated with Heckman's method and compared with ordinary least squares regressions uncorrected for sample selection bias. The IMR variable was estimated to have a negative effect on LOR, but the estimates were never highly significant or quantitatively important. Furthermore, bonus effects were virtually the same between methods. Tsui et al. (2005) reach a similar finding in their study of LOR and Simon et al. (forthcoming) find that sample selection bias corrections had little effect on estimates of the effects of GI Bill benefits on GI Bill use. Because sample selection bias corrections were found to have little influence on the estimates, we chose to correct for censoring by using Tobit estimation of the LOR equation and ignore correction for sample selection bias. This strategy is also followed in Chapter Eight, where LOR models are presented for all services.

Full LOR regression results are in Table C.3. Table 7.10 summarizes the estimates of interest, which are the effects of changes in the bonus multiplier on LOR. Each number in Table

---

[31] For a discussion of the Tobit method, see Cameron and Trivedi (2005), pp. 536–544.

Table 7.10
Regression Estimates of the Effects of the SRB Multiplier on Reenlistment Months, by Zone

| | SRB Multiplier Change | | | | | | |
Zone	0 to 0.5	0.5 to 1	1 to 1.5	1.5 to 2	2 to 2.5	2.5 to 3	3 to 3.5
	FY 2002–2006 Data						
A	4.5*	3.6*	1.4*	1.4*	–3.2*	–3.7*	–2.4*
B	10.5*	2.8*	–1.8*	–0.6	–5.0*	–2.3*	–2.5*
	FY 2002–2004 Data						
A	4.7*	2.1*	2.8*	4.4*	1.7	–4.8*	–0.8
B	10.2*	2.3*	2.3*	3.5*	–8.2*	–1.5	10.5*

NOTES: Full model estimates are provided in Table C.3. Estimates were obtained using the Tobit regression method and are based on the LOR and SRBM of actual reenlistees.

* Denotes statistical significance at the 5 percent level.

7.10 indicates the estimated change in months of reenlistment that results from an increase in the SRB multiplier from the next lower value.

The estimates in Table 7.10 give a similar picture as the means reported in Table 7.9. According to the estimates based on the full FY 2002–2006 period, multiplier increases are associated with LOR increases up to a multiplier of 2 (Zone A) and 1 (Zone B); beyond these values, LOR is estimated to fall as multipliers increase. Evidence that the negative association between multipliers and LOR may be due in part to bonus ceilings is provided by the lower-panel estimates in Table 7.10, which exclude FY 2005–2006 data. Here, SRB multiplier increases are associated with LOR increases up to a multiplier of 2.5 (Zone A) and 2 (Zone B). Beyond these values, LOR is estimated to remain the same or to decline as multipliers increase. Such declines are consistent with the fact that bonus ceilings are binding for most personnel when multipliers reach 2.5–3.[32]

The estimated effects of SRB multiplier changes on LOR are broadly similar to the effects estimated by Tsui et al. (2005) in an analysis of Army LOR and by Hogan and Simonson (2007) in an analysis of Navy LOR. Tsui et al. estimate that up to a multipler of 4, each one-level increase in the SRB multiplier raises the length of a reenlistment by about four months. According to their results, personnel eligible for a multiplier of 4 reenlist for almost 1.5 years longer than personnel not eligible for an SRB. They estimate that, beyond a multiplier of 4, multiplier increases cease to have an effect on LOR (but they do not have a negative effect). The result that LOR responds positively to an SRB multiplier increase when the SRB multiplier is either 0 or small to begin with is consistent with the findings in our study. However, the Tsui et al. study implies positive effects on LOR at higher multipliers and nonnegative effects at any values of the multiplier. Our analysis, by contrast, finds that multiplier increases appear to have negative effects beyond a multiplier of 2 at either zone.

Regression estimates in Table C.3 indicate that deployed personnel reenlist for about three months more than otherwise similar nondeployed personnel. It is interesting to note

---

[32] There is an inconsistency in Table 7.10 relating to the large positive estimated Zone B effect of going from a multiplier of 3 to a multiplier of 3.5 or above. This estimate, while statistically significant, should be downplayed. Only 23 Zone B reenlistees in the FY 2002–2004 period received an SRBM of 3.5 or above.

that deployed personnel who are subject to stop-loss reenlist for about four months more than deployed personnel who are not subject to stop-loss. Since those who are subject to stop-loss reenlist at a much lower rate than those who are not, perhaps the longer reenlistments among the stop-loss personnel may reflect a selection effect and not a causal effect of stop-loss status on reenlistment.[33] Personnel with more past deployment reenlist for shorter periods, but the differences in reenlistment months associated with past deployment differences are minor. Those with 13–24 months of past deployment reenlist for about two months less, on average, than those with no past deployment. Differences in reenlistment months associated with differences in most other variables are also minor. Most are on the order of three months or less.

The Cost-Effectiveness of Selective Reenlistment Bonuses

The estimates presented in the sections above indicate that Selective Reenlistment Bonuses affect both the rate of reenlistment and the length of reenlistment. This section explores the implications of the estimates for the cost-effectiveness of the program. A simple way to do this is to conduct a thought experiment in which the SRB program is eliminated. Since FY 2006 was a year with relatively large SRB payments and a large percentage of reenlistees received SRBs, we selected FY 2006 as the baseline for the thought experiment. The thought experiment is conducted within the context of the 24-MOS sample on which the empirical analysis was based.

Consider first the probability of reenlistment. The fitted annual data models for the probability of reenlistment reported in Table C.1 were used to predict the reenlistment probability of each individual in this dataset who was at ETS in FY 2006 using that individual's personal information, including SRB multiplier, where the multiplier is defined conditional on deployment status. Then each individual's reenlistment probability was predicted assuming a 0 SRB multiplier. By the same process, each reenlistee's length of reenlistment was predicted with and without the program, using the FY 2002–2006 regressions for LOR reported in Table C.3. It is useful to note that the average Zone A SRB multiplier in FY 2006 was 1.53, whereas the average Zone B multiplier was 1.29.[34]

Table 7.11 shows the results of this experiment for Zones A and B separately. Eliminating the SRB program entirely is predicted to reduce the probability of reenlistment in Zone A from 36 to 27.5 percent, or by 8.5 percentage points. Likewise, the Zone B reenlistment rate is predicted to fall by 7.9 percentage points.[35] Among those who reenlist, the average length of a Zone A reenlistment is predicted to decline from 42 months with the program to 36

[33] The significantly lower reenlistment rate of those subject to stop-loss indicates a lower average taste for military life among personnel subject to stop-loss. But stop-loss personnel who do reenlist may, therefore, have a higher average taste for military life than other personnel, who reenlist at higher rates. If so, they may be willing to reenlist for longer periods.

[34] These are unconditional multipliers—that is, they include 0 values for those not eligible for an SRB. The FY 2006 conditional averages at Zones A and B were 2.3 and 2.2, respectively.

[35] These effects are larger than the estimated effects of a one-multiple change in SRBM reported in Table 7.6 because the average value of SRBM is larger than 1 in both zones. Also, the estimated effects are those for the case where SRBM is defined conditional on deployment status. The estimated effects would be smaller if we used the estimates where the definition is not conditional on deployment status. Below, we use a smaller estimate, consistent with our lower-bound estimates in Table 7.6.

Table 7.11
Predicted Reenlistment Effects Resulting from Elimination of the SRB Program, by Zone, FY 2006

Zone	Predicted Percentage Reenlisting With SRB	Predicted Percentage Reenlisting Without SRB	Change in Percentage Reenlisting	Predicted Average LOR With SRB	Predicted Average LOR Without SRB	Change in Average LOR
A	36.0	27.5	–8.5	42	36	–6
B	50.4	42.5	–7.9	45	38	–7

months without the program. In Zone B, the average LOR is predicted to decline from 45 to 38 months.[36]

Table 7.12 demonstrates the cost implications of the SRB program per 100 soldiers eligible to reenlist. With the SRB program, if 100 soldiers are eligible to reenlist in Zone A and 36 reenlist for an average period of 42 months, or 3.5 years, the program will yield 36 × 42/12 = 126 person-years of service in the second term of service. Without the program, there are predicted to be only 27.5 × 36/12 = 82.5 person-years. In FY 2006, the average monthly basic pay of the Zone A personnel at ETS was $2,006; in Zone B, the average was $2,358. The average months of reenlistment of all Zone A reenlistees was 41, whereas the average in Zone B was 44. Using an average Zone A multiplier of 1.53, the average Zone A bonus payment was 1.53 × (41/12) × $2,006 = $10,486. The Zone B average payment is 1.29 × (44/12) × $2,353 = $11,153.

With 36 reenlistments in Zone A receiving an average payment amount of $10,486, the Zone A SRB budget is $377,496 (per 100 personnel eligible to reenlist). Likewise, with 50.4 reenlistments in Zone B at an average payment amount of $11,153, the Zone B SRB budget is $562,111 (per 100 personnel eligible to reenlist). Dividing these budget amounts by the number of person-years gives the SRB cost per person-year. The Zone A and Zone B costs are $2,996 and $2,974, respectively. The SRB cost per additional person-year produced by the program (i.e., the marginal cost) is obtained by dividing the SRB budget by the difference in person-years with and without the program. The calculated amounts are $8,678 in Zone A and $15,090 in Zone B, respectively. It is interesting to note that in both zones, the SRB cost of an additional person-year is about 36 percent of the average annual basic pay of the personnel at ETS in FY 2006.

Table 7.12 reveals a fact of life about the military compensation program in general and the SRB program in particular: The marginal person-year cost of compensation system changes is much higher than the average cost. Marginal cost exceeds average cost, because many personnel choose to remain in the military even after pay is reduced.[37] In the case studied here, elimination of SRBs altogether reduces reenlistment somewhat but does not eliminate it altogether. Looked at the other way, because the extra person-years induced by an SRB are less than the person-

[36] The predicted LOR was close to the averages observed in the data. Among the Zone A personnel who received SRBs in FY 2006, the average reenlistment was 42 months, whereas the average LOR among those not receiving SRBs was 37 months. In Zone B, the averages were 47 and 37, respectively.

[37] This discussion implies the payment of "economic rent." Economic rent to labor is payment above that necessary to induce a given supply of labor (i.e., payment above opportunity cost). Military personnel who would have reenlisted in the absence of an SRB earn economic rent when an SRB is implemented. Economic rent varies inversely with the responsiveness to SRB, because the lower the responsiveness, the higher SRBs must be set to obtain a given number of reenlistments.

Table 7.12
SRB Program Costs per 100 Soldiers Eligible to Reenlist Using Annual Data Estimates, by Zone

Zone	Person-Years With SRB	Person-Years Without SRB	Person-Year Change	Average SRB Payment	SRB Budget	SRB Cost per Person-Year	SRB Cost per Additional Person-Year
A	126	83	43	$10,486	$377,496	$2,996	$8,678
B	189	135	54	$11,153	$562,111	$2,974	$10,409

NOTE: Amounts are in FY 2006 dollars.

years obtained without the SRB, the average SRB per person-year (after implementation) must be smaller than the per-year cost of the extra years induced by the SRB (i.e., the marginal cost).

Table 7.12 shows that the cost of an additional person-year is about $1,800 higher in Zone B than in Zone A. Part of this cost difference is due to the higher basic pay of the personnel in Zone B ($2,356 versus $2,006). Another part of the cost difference stems from the fact that the reenlistment rate that would prevail in the absence of an SRB is higher in Zone B than in Zone A (42.5 percent versus 27.5 percent). The higher the retention rate in the absence of a compensation increase, the greater the marginal cost of the compensation increase inclusive of economic rent. An offsetting factor that reduces the cost difference between the zones is that reenlistments are longer in Zone B than in Zone A (Table 7.11).

The marginal SRB cost estimates in Table 7.12 are based on the annual data model estimates of the effect of SRB multiplier changes on the probability of reenlistment where the multiplier is defined conditional on deployment. The annual data model estimates are somewhat higher than those obtained using monthly data found in Table 7.8 above, and these estimates are higher than those obtained using annual data but where the multiplier is not conditional on deployment. According to the monthly estimates, a one-level SRB multiplier decrease reduces Zone A reenlistments by about 4 per 100 individuals eligible for reenlistment and Zone B reenlistments by about 3 per 100 individuals eligible for reenlistment. Past studies provide estimates in the range of 2–6 reenlistments per 100 eligible, with a median estimate of 3, consistent with the estimates in Table 7.6 for the case where the bonus multiplier is not conditional on deployment status. To see the cost implications of these lower estimates of the responsiveness of reenlistment rates to SRB multiplier changes, Scenario A in Table 7.13 repeats the marginal cost calculations found in Table 7.12 assuming that each one-multiplier SRB decrease reduces the probability of reenlistment by 3 percentage points. But Scenario A continues to use the estimates of the effects of multiplier changes on the length of reenlistment found in Table 7.12.

Scenario A may overstate the person-year gain to reenlistment that results from longer reenlistments. Since many personnel who reenlist at the end of one term are going to reenlist at the end of the next term anyway, the person-year gain to longer reenlistments comes only from personnel who are going to leave at the end of their reenlistment terms. About 50 percent of Zone A reenlistees reenlist in Zone B and about 70 percent of Zone B reenlistees reenlist in Zone C. Scenario B thus assumes that the Zone A LOR gain from higher SRB (or loss from lower SRB) is only half that in Scenario A and that the Zone B LOR gain (or loss) is only 30 percent of that in Scenario A.

In Scenario A, the SRB cost of an additional person-year is $11,870 in Zone A and $13,383 in Zone B, about one-third higher than previous estimates. In Scenario B, where personnel are the least responsive to SRB changes, marginal SRB cost is estimated to be $15,729

Table 7.13
SRB Cost per Additional Person-Year Using Two Alternative Scenarios, by Zone

Zone	Alternative Scenario A		Alternative Scenario B	
	Person-Year Change	SRB Cost per Additional Person-Year	Person-Year Change	SRB Cost per Additional Person-Year
A	32	$11,870	24	$15,729
B	42	$13,383	26	$21,620

NOTES: Scenario A assumes (a) that each one-level change in the SRB multiplier changes reenlistments by 3 per 100 eligible to reenlist and (b) that the LOR response is the same as in Table 7.12. Scenario B assumes the same reduced reenlistment responsiveness as in Scenario A and further assumes one-half of the Zone A LOR responsiveness of Scenario A and only 30 percent of the Zone B LOR responsiveness of Scenario A. Amounts are in FY 2006 dollars.

for Zone A personnel and $21,620 for Zone B personnel. The latter marginal cost estimates, especially, are very high compared to the average person-year cost of SRBs of about $3,000 (Table 7.12). How cost-effective are SRBs? As discussed above, SRBs can be targeted to specific skills experiencing retention problems without the need to change compensation for skills not experiencing retention problems. They are clearly more cost-effective than general pay increases, which cannot be targeted at specific skills or specific career points, as SRBs can be.

Finally, it is useful to compare the marginal cost of person-years obtained with SRBs to the marginal cost of person-years obtained via initial enlistment bonuses. Chapter Four estimated the cost of an additional high-quality enlistment if the enlistment bonus were $44,900. Assuming a four-year enlistment, the implied person-year cost is $11,225. This cost is not much different from the Scenario A marginal SRB costs in either Zone A or Zone B but is considerably lower than our more pessimistic Scenario B estimates of $15,729 and $21,620, respectively. In reality, bonus dollars spent on reenlistments are more cost-effective than bonus dollars spent on initial enlistments, for three reasons. First, reenlistees do not require training, but new enlistees do. Second, studies show that experienced personnel are more productive than first-term personnel (see Warner and Asch, 1995, and Asch, Hosek, and Warner, 2007, for reviews of the evidence). Third, non-ETS attrition beyond the initial enlistment is typically very low, on the order of 3–5 percent per year. Although our marginal cost estimates for reenlistment bonuses do not account for non-ETS attrition, accounting for it would not change them much. On the other hand, Chapter Five showed that first-term attrition is much higher. Roughly one-third of high-quality enlistments fail to complete the initial enlistment and much of the attrition occurs in the first two years, often before personnel have been trained and become productive to the Army. That is to say, the marginal cost of a productive person-year via initial enlistment bonuses is much higher than $11,225. As a result, bonus dollars focused on reenlistments may be more cost-effective than bonus dollars focused on initial enlistments.

Concluding Remarks

The wars in Afghanistan and Iraq have placed great stress on the U.S. armed forces, leading to significant challenges in both recruiting and retention. The Army imposed a stop-loss policy on a significant fraction of its enlisted forces and, indeed, soldiers who received stop-loss orders were less likely to reenlist. We do not wish to understate the policy significance of stop-loss; nevertheless, it is important to note that the reenlistment rate of personnel in Zone A under stop-loss was about two-thirds that of soldiers who were not under stop-loss. In other words, only a third of the soldiers under stop-loss would have left military service if given the opportunity to do so.

We found that longer past cumulative deployment was associated with lower rates of reenlistment. However, we found that soldiers currently deployed had *higher* rates of reenlistment. The question naturally arises whether this effect is causal, for there are a number of possible interpretations. The most straightforward (compelling) explanation is that soldiers time their reenlistments to coincide with deployment because bonuses received while in a combat zone are not taxed; the higher rate of reenlistment among those who are deployed may merely reflect this tax advantage. There are other possible explanations. One is that there is a camaraderie effect in which soldiers in a unit deployed in time of war are more willing to serve to not "let down" their fellow soldiers. Another arises from Army rotation policy. Army policy is to rotate soldiers out of the theater of war after a single tour of duty. Soldiers may prefer not to be deployed, with those who are currently deployed expecting not to be deployed in the near future and therefore reenlisting at a higher rate and the opposite being true for those who are currently not deployed. Distinguishing between these explanations appears to be an important avenue for future research.

Finally, our analysis indicates significant positive effects of selective reenlistment bonuses on the likelihood of reenlistment and on the length of reenlistment, particularly at the first reenlistment point. Calculations based on the estimates indicate that the cost of an additional person-year induced by the selective reenlistment bonus is around $12,000 to $20,000, depending on reenlistment zone and on which estimates of bonus responsiveness are used. Although not low by any measure, these costs compare favorably with the cost of alternative policies to induce additional person-years. Higher SRB multipliers result in longer lengths of reenlistment in Zone A, up to a point. However, we find that at high multipliers, the effect of a higher SRBM is negative. This negative effect is more pronounced for reenlistments in Zone B. The negative SRBM effect on length of reenlistment may be due to diminishing SRBM effects on term length at higher SRBMs, because people are willing to trade off a higher SRBM for a shorter term, or it may be due to the constraining presence of bonus caps that eliminate the incentive to choose a longer term. Nonetheless, the evidence presented in this chapter suggests that selective reenlistment bonuses pay a vital—indeed essential—role in Army force management.

CHAPTER EIGHT

Reenlistment Results for All Services

The previous chapter focused on reenlistment in the Army. We take advantage of recently published results reported in Hosek and Martorell (2009; referred to henceforth as HM) to present results in this chapter from reenlistment models estimated for all services. The approach in this chapter largely parallels that of the previous chapter and, at the same time, the approach extends the analysis reported in HM. The HM study focuses on the extent to which deployments during the global war on terrorism (GWOT) have changed the willingness to reenlist. HM also discusses the role played by reenlistment bonuses. In particular, it estimates the effect of SRBs on reenlistment and uses these estimates to calculate the importance of increases in SRBs on maintaining reenlistment rates in recent years.

The analysis in this chapter is presented separately from the results for the subset of Army occupations from Chapter Seven for two reasons. First, Chapter Seven contains more detail about the particulars of the Army's bonus-setting policies, which may be useful for readers or policymakers interested in how the Army managed reenlistment bonuses during the GWOT. Second, as we explain below, the statistical methodology in this chapter is somewhat different from that in the preceding chapter; comparing the estimates will be informative about how sensitive the estimates are to the statistical tools used.

Although the results in this chapter are based on an approach similar to that in the previous chapter, there are differences in data and method. This is a consequence of the HM study's being done before, and independently of, the analysis in Chapter Seven. However, the work presented here builds on the HM study in ways that make it consistent with the approach in the preceding chapter. First, we use a more refined SRB measure than in HM—one that more accurately reflects what service members would face when making their reenlistment decisions. Second, we examine the effect of the bonus on the length of reenlistment, doing so with the same specification of the SRB multiplier variable as in Chapter Seven. Third, in keeping with Chapter Seven, we present estimates of the additional reenlistments in the period January–September 2007 that were generated by an increase in the SRB multiplier from a value of 0 to its value in FY 2007, and we relate these to the cost of the bonus program.

We first describe the data and econometric strategy that were used in HM. The HM database covers all four services and was created separately from that in the previous chapter, although both draw on Defense Manpower Data Center data. Then, we discuss trends in bonus generosity during the period FY 1996–2007 and present the results from the reenlistment and length of reenlistment models. Finally, we present estimates of the additional years of service induced by the bonuses in FY 2007, relative to no bonuses, and the cost associated with this. Along the way, we compare our methodology and results for the Army to those presented in Chapter Seven, as appropriate.

Data and Econometric Strategy

Sources of Data

The DMDC Proxy PERSTEMPO file was our main data source on deployments, reenlistment decisions, and service member characteristics. This file consists of monthly records of all active duty service members and has information on military occupational specialty, marital and dependents status, educational attainment, pay grade, measures of deployment, and the time remaining on the current term of service. The PERSTEMPO file also specifies gender, date of birth, race, ethnicity, and Armed Forces Qualification Test category. To this file, we merged information from the DMDC Joint Uniform Military Pay System file, which contains disaggregated information on the types of pays that service members receive each month, including whether a member received a reenlistment bonus. The records indicate the size of the total reenlistment bonus as well as the amount paid in a particular month. We use the total amount of the bonus and the member's pay grade to infer the bonus multiple (or "step"), which indicates the amount of the bonus offered. The bonus multiple reflects the generosity of the bonus as determined by bonus-setters operating under service and OSD bonus policy and, as such, the bonus multiple is the key bonus variable used in the analysis of the effect of bonuses on reenlistment. In contrast, the amount of the individual's bonus would not be appropriate as an explanatory variable, because it reflects both bonus generosity and the length of reenlistment, the latter being chosen by the service member.

With respect to the bonus history presented in Chapter Six, the Army began implementing the Enhanced SRB in June 2007, three months before the end of our data window, September 2007. Enhanced bonus amounts are based on a table rather than on a single formula as used for the SRB. Use of data on the total amount of the bonus and the member's pay grade allows us to infer an implicit bonus multiple. This method of gauging the generosity of the enhanced bonus helps to place it on the same basis as that for the SRB.

Identifying Reenlistment Decisions

As in Chapter Seven, HM infers reenlistment decisions from monthly information about ETS. We briefly describe this algorithm, which was first used in Hosek and Totten (2002).[1]

To identify decisions, individuals are followed over time in the PERSTEMPO data. As the ETS date approaches, members make one of three decisions. They can reenlist, exit the military, or extend their contract. Extensions and reenlistments differ in a qualitative sense. Extension is a way to postpone the reenlistment decision if the individual is uncertain or prefers to stay in the military a short time longer, e.g., to continue to assist the unit in completing a task or to allow a son or daughter to complete the school year. Also, extensions might be used to time a reenlistment so that it occurs while the service member is deployed, at which point any reenlistment bonus would be subject to preferential tax treatment if the deployment is to a combat-zone tax-exclusion area. We identify exits by observing when the ETS falls to zero and the pay grade becomes missing (indicating no further receipt of income from the military). Reenlistments are identified as an increase in ETS of 24 months or more, with the length of reenlistment defined as the net change in ETS.[2] We define an extension as an increase in

---

[1] See Hosek and Totten (2002) for a more thorough discussion.

[2] Although we view a 24-month cutoff as reasonable, individuals may not be eligible for SRBs unless they reenlist for 36 months or more. Thus, SRBs may have a weaker effect on reenlistments of between 24 and 35 months than on reenlistments of 36 months or greater.

months to ETS of 23 or fewer. Here, we focus on whether someone exits the military or reenlists. Therefore, if a member extends, he or she is not considered to have made a reenlistment decision and is instead followed until exit or reenlistment.

The same approach applies to service members under stop-loss. A service member under stop-loss may not leave the military, which implies that during stop-loss no exit can occur (except for serious injury or death). However, service members under stop-loss can reenlist or extend, so such decisions can be observed. A service member under stop-loss who wishes to reenlist may do so and therefore is not constrained by the stop-loss policy. But a service member who wishes to leave is constrained and must postpone his or her decision. In our approach, such individuals are followed until a stay/leave decision is observed, which would be after the stop-loss stricture. Although the PERSTEMPO file does not contain a flag for stop-loss, HM presents indirect estimates of the number of service members constrained by stop-loss. The estimates for the Army at first-term reenlistment are 2.8 percent in FY 2003, 6.7 percent in FY 2004, 10.0 percent in FY 2005, 4.4 percent in FY 2006, and 3.6 percent in FY 2007. The estimates for Army second-term reenlistment are lower. Thus, the estimates of constrained personnel are about one-third of the estimates of personnel deployed under stop-loss shown in Chapter Seven. The HM estimates for the other services show lower prevalence of stop-loss except for second-term airmen, where the percentage is similar to that of the Army second-term, namely, about 4 percent.

Our sample and approach to identifying reenlistment decisions includes, for example, service members who made a reenlistment decision at ETS as well as those who did so at, say, 12 months before ETS. As explained next, the bonus information we attach to the individual's record is the bonus multiple for the month in which the individual makes the reenlistment decision.

Reenlistment Bonuses

For the estimates in Chapter Seven and in this chapter, we did not have access to SRBM levels officially set by the services. Instead, we use an empirical approach to approximate the SRBM an individual would have faced when making the reenlistment decision. The first step was to identify individuals who reenlisted and who received a bonus. It is this group for whom the bonus is observed; it is not observed for the leavers. We infer the bonus multiplier from those who reenlist and impute the inferred multiple to all individuals making a reenlist/leave decision at this time. The multiplier associated with this bonus was calculated by the formula used to determine the total SRB: the product of term length (the years of additional service obligated by the new contract), monthly basic pay at the time of reenlistment, and the bonus multiplier. For individuals who reenlisted but did not receive a bonus, the multiplier was set to 0. To calculate the SRBM available at the time of the reenlistment decision, HM took the average multiplier in cells determined by zone, three-digit DoD duty MOS, and quarter of the year. This value was then assigned to all individuals in the cell making a decision (including those who did not reenlist).

An important contribution of the current work is a refinement of the bonus measure used in HM. The refinement makes the construction of the bonus variable similar to, although not identical with, the approach in Chapter Seven. One refinement is to use service duty occupation rather than DoD occupation, because bonuses are set at the service occupation level, and the DoD occupation code sometimes combines several service occupations. Second, we use pay grade to create the cells, since bonuses can differ by pay grade within an occupation-zone cell. Third, since bonuses can be reset more often than at the quarterly level, we further define cells by the month of the reenlistment decision rather than the quarter. Fourth, we rescale the SRBM to account for growth in real basic pay over the sample period. Growth in real basic pay amplifies the effect of a one-unit change in the multiplier on the size of a total bonus. Targeted pay increases led to increases in real basic pay of 15 to 24 percent for E-3 to E-6, depending on year of service, from FY 1996 to FY 2007. However, much of the increase was accomplished in FY 2000, and the increases in real basic pay from FY 2002 to FY 2007 were only 2 to 6 percent. We scaled the SRBM so that a one-unit change in the nominal multiplier has the same value in all years. In other words, if real basic pay was, say, 4 percent higher in one year than in the base year, we rescaled the SRBM so that it was 4 percent higher in that year.

Finally, as discussed in Chapter Seven, the SRBM used in this chapter is calculated in two ways: (a) cells defined by zone, three-digit service duty MOS, pay grade, and time period (month of reenlistment decision); and (b) the same plus whether the service member was deployed. In this chapter, deployment status is defined on the basis of whether the individual was deployed in the month of the reenlistment decision.[3] We do this because, as described in Chapter Six, the services offered a deployment reenlistment bonus to service members who reenlisted when deployed and were not in a specialty that offered an SRB. However, as described in Chapter Seven and reiterated below, allowing the SRBM to vary by deployment potentially introduces other sources of bias.

Creation of the Analysis File

We focus on first and second reenlistments. To create the analysis file used in the estimation, we searched through the PERSTEMPO file for instances of reenlistments or exits and constructed an analytic file consisting of records of these decisions. Individuals who reenlist at the end of their first term may show up once again in the file if their second-term reenlistment decision is observed before the end of the data (September 2007). To focus on the GWOT period, our analysis sample begins in January 2002 (see HM for results from the period FY 1996–2007). Most individuals who reenlist twice become career military personnel.

In addition, we made three other sample restrictions. First, missing data: Individuals with missing data for AFQT category, race, average SRBM, or educational attainment were dropped. Note that, as described above, individuals will have missing SRBM values if they are in a cell with no reenlistments. Second, attritors: Members who exited the military more than six months from the end of their ETS were excluded.[4] Most attrition occurs within the first year of service; exiting early is often a sign of an involuntary separation or a separation that is due to unusual or extenuating circumstances unlikely to be driven by deployment experiences.[5] Early exit can also result from disability, which may be related to injuries sustained on deploy-

---

[3] More precisely, we check whether the individual was deployed in the month of the reenlistment decision or in the month before or the month after.

[4] Only those who separate more than six months from the end of their ETS are excluded. Those who reenlist more than six months from the end of their ETS are included. The reason for the asymmetric treatment is that the decisions are asymmetric; someone can reenlist early, but they cannot exit early, at least not voluntarily under normal circumstances.

[5] If service members exited early to avoid deployment, we would expect to see an increase in early exits starting in FY 2002. To investigate this possibility, we examined DMDC data on enlisted continuation rates by service for year 1 to year 2 and for year 2 to year 3. There is no decrease in continuation in any of the services, with the exception of the Air Force in FY 2004 and FY 2005, where the continuation rate decreased from a high of 93 percent in FY 2003 to a low of 89 percent in FY 2005. The other services had fairly stable continuation rates, as did the Air Force in other years.

ment. Finally individuals were excluded if they had completed fewer than three years of service at the time of their reenlistment decision. This restriction was made because members with fewer than three years at the time of their decision typically signed an initial service contract for two years (although it also includes individuals who signed a longer service contract but reenlisted early).

Econometric Issues

We use a linear probability model to estimate the effect of the SRBM on the probability of reenlistment and a Tobit model to estimate the effect of the SRBM on the length of reenlistment. The linear probability model is easy to interpret and performs well in comparison to a probit or logit model, given that the probability of reenlistment is in the linear range of the error distribution (Wooldridge, 2001). The Tobit specification accounts for the fact that the LOR is censored, as mentioned in Chapter Seven. Those who exit the military have an LOR of zero, and those who reenlist have an LOR of at least 24 months, by definition. Also, the Tobit results can be used to infer the probability of reenlistment. However, using the Tobit in this way assumes that the regression coefficients for the reenlistment margin are the same as those for the LOR margin, whereas estimating a separate reenlistment model does not. A potential advantage of analyzing reenlistment and LOR separately is that policymakers may care more about filling spaces in the immediate term, in which case the reenlistment response may be more important than the LOR response conditional on reenlisting. Another econometric consideration is that using observations for those who reenlist could lead to biased LOR estimates if selection into reenlistment is correlated with LOR. For instance, individuals who face a small SRBM yet decide to reenlist may have a high taste for the military, leading them to reenlist for a relatively long time. This bias, resulting from a positive correlation between unobserved factors in the reenlistment and LOR equations, would bias the bonus estimate downward in the LOR equation. But this bias is likely to be small; as discussed in Chapter Seven, models controlling for error correlation gave estimates that were much the same as models that did not. The results we report are from models that do not allow for error correlation.

Estimation of the effect of SRBMs is complicated by reverse causality: The services adjust bonuses to attain their reenlistment targets and maintain a steady flow of personnel, by occupation, into higher years of experience and higher grades. Bonuses are driven to some extent by persistent attributes of an occupation, such as the type of assignments, tasks, work conditions, and deployment, as well as by the civilian job opportunities most relevant to the occupation. Bonuses help to adjust for these persistent differences, e.g., higher bonuses for less-attractive specialties, specialties with a high replacement cost or high internal value (a critical skill), or attractive civilian opportunities. Bonuses thus serve to some extent as a compensating differential, bringing the expected reenlistment rate in a specialty to the rate desired by the service for force management objectives. We use occupation-specific fixed effects to account for these persistent differences.[6] Because general conditions affecting reenlistment in all specialties can vary over time, we also use fixed effects for year of decision. The occupation and year fixed effects account for permanent differences across occupations and differences across years in the

---

[6] This point is known from previous work; see the survey by Goldberg (2001). Hattiangadi et al. (2004) also use within-occupation variation in reenlistment bonuses to identify the effect of bonuses on reenlistment in the Marine Corps. Hansen and Wenger (2002, 2005) use occupation groups but also report results for specifications with fixed effects for occupational specialties (ratings) in the Navy.

need for bonuses. We expect these persistent and temporal factors to be systematically related to bonus-setting, which means that they are major sources of the reverse causality we seek to avoid in estimating the effect of bonuses on reenlistment. Given the fixed effects, the remaining bonus variation represents variation within an occupation over time. Still, this variation is not exempt from reverse causality.

When considering estimates from these models, the source of within-occupation variation in the SRBM matters. There are two reasons a bonus-setter would change the bonus in an occupation. First is a change in demand for personnel in that occupation. The bonus-setter will increase the bonus if demand increases. Related to this, the bonus budget might change for reasons outside the bonus-setters control, perhaps as a result of legislative forces, and bonus managers might change bonuses in response. Second is a change in the supply of personnel willing to reenlist. An increase in civilian-sector demand could cause the bonus-setter to increase the bonus. Appendix A in HM describes models of bonus-setting and discusses the possible biases in estimates of the bonus effect.

It is likely that changes in occupation demand are unrelated to the willingness of service members to reenlist in the occupation and, if so, demand-induced changes in bonuses can be used to identify the bonus effect without bias. But supply shocks are likely to be related to service members' willingness to reenlist, and bonus changes in response to supply shocks will cause bias in the estimated bonus effect. A decrease in reenlistment, other things equal, leads the bonus-setter to increase the bonus and produces a downward bias in the bonus coefficient.

Thus, the credibility of the estimated bonus effect on reenlistment rests on whether the within-occupation changes in bonus are in response to demand or supply shocks. But the data do not identify either type of shock, and there is no basis on which to dismiss supply shocks. Therefore, controlling for occupation and year fixed effects can be expected to lessen but not eliminate the possible downward bias in bonus estimates (see HM for a formalization of this point).

Finally, another source of bias may be present. Suppose a service is increasing or decreasing its size selectively by occupation, as might be expected if changes in force structure accompany changes in force size. In addition to using bonuses to expand certain occupations, the service might reallocate effort by career counselors away from occupations that are not growing and toward occupations that are growing, reallocate incentives such as choice of location or preference for further training in the same way, and constrain the number of positions open in occupations that are shrinking, creating a demand constraint (i.e., more service members want to reenlist that can do so). Such actions are correlated with bonus use but are not included in our data. The result is an omitted variable bias causing an upward bias in the bonus effect. The Navy and Air Force were downsizing and the Army and Marine Corps were growing in our data period. These changes might have led to an upward bias, a possibility that should be kept in mind along with the downward bias described above.

We use two alternative definitions of the bonus variable (as does Chapter Seven). These methods are of importance to the Army. As will be seen, Army bonus estimates depend on the method, but the estimates for the other services do not. One method defines the SRBM as the average over cells, depending on whether the service member was deployed at the time of reenlistment. This approach allows for different bonus levels depending on deployment status, but in assigning the average bonus to service members, it creates a measurement error because the average is not accurate for any given individual. The chief source of the error is the payment of so-called deployment bonuses. As noted in Chapter Six, a deployment bonus may be paid to individuals who reenlist when deployed and are not eligible for an SRB. The amount is averaged into the bonus for the specialty, whereas it is payable only to deployed members of the specialty. The introduction of this error can be expected to bias the estimated bonus effect toward zero.

The second method breaks apart the average and defines the SRBM conditional on deployment. In this approach, the average bonus in an occupation is computed for those who are deployed when they reenlist and separately for those who are not. The former is assigned to deployed service members at the reenlistment point in the occupation and the latter to nondeployed service at the reenlistment point in the same occupation. The problem with this approach arises mainly in the Army and comes from the way stop-loss constrains a service member's choices: Under stop-loss a deployed soldier can choose to reenlist but cannot choose to leave. So whenever we observe a deployment bonus, we expect to see only reenlistments, and the data largely confirm this. Among first-term soldiers making a stay/leave decision when deployed, 93 percent reenlist. This implies that conditioning the SRBM on deployment will likely cause an upward bias in the estimated bonus effect. The first method, which defines the bonus variable only in terms of occupational specialty, rank, and term, helps to "bury" this upward bias.

Two further theoretical issues concern estimation of the LOR response. Unlike the reenlistment response to an increase in the SRBM, which theory predicts will always be nonnegative, the prediction regarding the LOR response is not clear. One reason is that an increase in the SRBM generates an income effect, and if the option value of being able to leave the military sooner than otherwise is high enough, the service member may choose a shorter LOR when the SRBM increases. Suppose a 48-month reenlistment would have been chosen at the initial SRBM. An increase in the SRBM would increase income, and by choosing a somewhat shorter reenlistment, the service member may have both higher income and the liberty to leave sooner. The service member gains the opportunity to take advantage of attractive military or civilian offers that might arise in the months that are no longer committed to the military. Second, caps on bonus amounts can create incentives to reduce LOR when the SRBM increases.[7] Consider someone who would reenlist for 48 months and receive a bonus equal to the maximum SRB (or close to it) with the existing SRBM. If the bonus multiplier increases but the maximum does not, the individual gains nothing by choosing a longer LOR and, in fact, will be able to maintain his or her bonus at the maximum level at a shorter LOR. How relevant this is empirically is difficult to determine, because it requires knowledge of the SRB caps that the services set, which may vary by MOS and be limited overall by caps set by legislation. Bonus caps by specialty are not available in our data but, as with income effects, they may help to explain why a higher bonus may lead to a decrease rather than an increase in length of reenlistment.

Another aspect to consider is that a higher SRBM may induce service members with a relatively low taste for military service to reenlist. A higher SRBM might increase reenlistment but bring in those who choose a shorter LOR. Depending on the increase in reenlistment and the taste for military service among those reenlisting, this might reduce or even make negative the SRBM effect on LOR. However, we expect this selection effect to be small and to have little effect on the LOR bonus effect. The reason is that bonuses typically raise reenlistment by

---

[7] The caps could also attenuate the estimates of the reenlistment margin response, because the stated multiplier would exceed the effective multiplier in the presence of a binding cap. However, we expect this type of bias to be relatively small.

no more than a few percentage points relative to a base of 40-55 percent reenlisting at the first term and 60 percent or more at the second term.

Reenlistment Bonus Prevalence and Generosity

Figures 8.1–8.4 (drawn from HM) show first-term bonus use by service, before and during the OEF/OIF years (see Appendix E for average first- and second-term bonuses by year and occupation). Bonus use is measured by the percentage of personnel reenlisting at the first term who received a bonus and the average bonus step for those receiving a bonus. The Army and Marine Corps increased both the number and size of reenlistment bonuses in FY 2005–2007 compared with FY 2002–2004, and the increases were substantial.[8] The percentage of soldiers reenlisting with a bonus decreased from between 43 and 53 percent in FY 1999–2002 to 16 percent in FY 2003 and FY 2004, then increased to 71 percent in FY 2005 and 79 percent in FY 2006 and FY 2007 (left panel of Figure 8.1). The large expansion in FY 2005 may have been driven by the Army's plans to increase personnel strength, enacted in FY 2004, rather than by looming difficulties with reenlistment from deployment. The Army's average bonus step for occupations that offered a bonus stood at 1.6 in FY 2001 and FY 2002, fell to 1.3 in FY 2003 and FY 2004, and increased to 1.8 in FY 2005, 2.2 in FY 2006, and 2.1 in FY 2007 (right panel in Figure 8.1). As a result of these changes, the percentage of reenlisting soldiers who received a bonus increased more than fourfold between FY 2003 and FY 2005–2007, and the average generosity of the bonus increased by more than 50 percent. The percentage of

Figure 8.1
Reenlistment Bonus Prevalence and Average Step, Army First Term

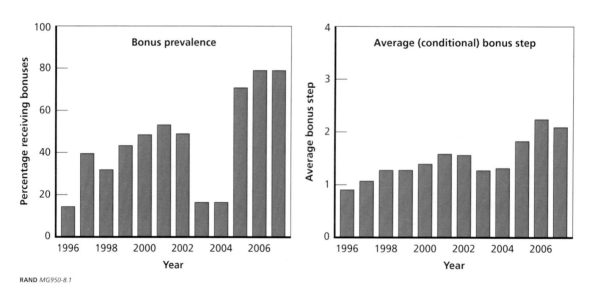

RAND *MG950-8.1*

---

[8] Note that in contrast to the results in Chapter Seven, these figures include all Army MOSs, so the magnitudes and timing of changes in the bonus generosity or incidence will be different from what is reported in Chapter Seven (Figure 7.2). Also, here we show bonus prevalence and the conditional-on-bonus-receipt average bonus multiplier, whereas Chapter Seven shows the unconditional multiplier (i.e., includes zeroes for service members who did not receive a bonus).

Figure 8.2
Reenlistment Bonus Prevalence and Average Step, Marine Corps First Term

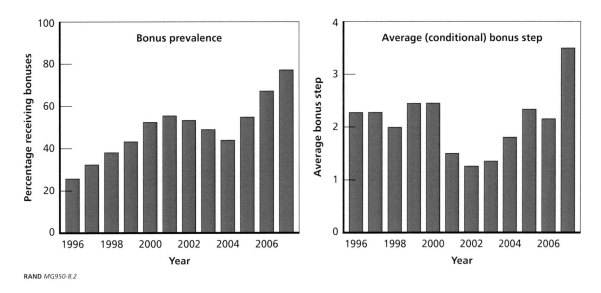

Figure 8.3
Reenlistment Bonus Prevalence and Average Step, Navy First Term

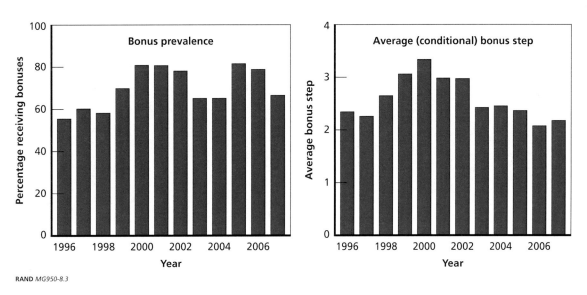

marines receiving a bonus increased from 43 percent in FY 2004 to 78 percent in FY 2007, and the average step climbed from 1.3 in FY 2002–2003 to 1.8 in FY 2004, 2.2–2.3 in FY 2005–2006, and 3.5 in FY 2007. In contrast, the percentage of seamen receiving a bonus held fairly steady at 70 to 80 percent, and the average step decreased from about 2.4 in FY 2003–2005 to 2.0 in FY 2006–2007. The Air Force decreased the percentage receiving a bonus from more than 80 percent in FY 2002–2003 to 14 percent in FY 2006–2007, and the average step increased from 3.2–3.4 in FY 2003–2004 to 3.7 in FY 2006–2007, a roughly 12 percent increase.

Figure 8.4
Reenlistment Bonus Prevalence and Average Step, Air Force First Term

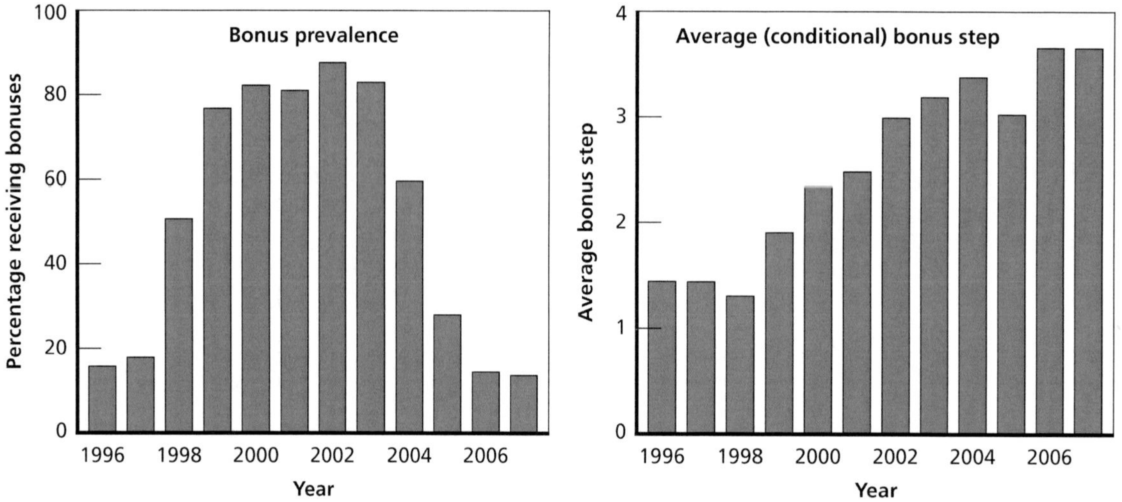

RAND *MG950-8.4*

As the evidence indicates, the branches with the heaviest combat duties in OEF and OIF had the largest increases in bonus use and generosity. These branches—the Army and Marine Corps—also embarked on efforts to grow in FY 2005. Bonuses helped to support growth and compensate for the heavy deployment.

Results

Table 8.1 shows the estimated effect of the SRBM on the probability of reenlistment. The estimates are obtained from linear probability models, where the dependent variable is an indicator for reenlistment and the key explanatory variable is the SRBM facing an individual at the time of the reenlistment decision (the notes to Table 8.1 list other variables included in the model). Results are presented by service for first- and second-term reenlistment. We report results from specifications where the SRBM variable for an occupation in a given month is an average across deployed/nondeployed statuses and where it is conditional on deployment status. The estimated coefficients for other variables included in the model, such as the deployment measure, are reported in Appendix D.

Starting with the results where the SRBM is not conditioned on deployment, all the bonus estimates are positive except for second-term Marines, where the coefficient is negative and not statistically different from zero. As noted in Chapter Seven, most estimates of the effect of the SRBM in the past literature range from 0.02 to 0.06. Our first-term Army, Navy, and Marine Corps estimates are in this range, and the Air Force estimate is 0.016. The second-term Army estimate is also in this range but the other second-term estimates are smaller.[9]

---

[9] These results are also different from those in HM for these years. Using the refined bonus measure leads to an increase in the estimated effects for first- and second-term Army personnel (from 0.013 to 0.025 for the first term, and 0.016 to 0.025 for the second term). For the Marine Corps and Navy, however, the bonus estimates in Table 8.1 are lower than those reported in HM. The estimates for the Air Force are not substantially different.

Table 8.1
Effect of the SRBM on the Probability of Reenlistment, by Service,
FY 2002 to September 2007

| | SRBM Varies by Deployment? | | | |
| | First Term | | Second Term | |
	No	Yes	No	Yes
Army	0.025**	0.089**	0.025**	0.051**
	(0.002)	(0.002)	(0.002)	(0.002)
	129,322	123,774	85,969	84,055
Navy	0.025**	0.025**	0.010**	0.009**
	(0.002)	(0.002)	(0.002)	(0.001)
	96,334	93,949	58,107	57,104
Marine Corps	0.036**	0.035**	−0.004	−0.003
	(0.002)	(0.002)	(0.004)	(0.004)
	77,214	73,744	22,489	22,143
Air Force	0.016**	0.013**	0.015**	0.014**
	(0.002)	(0.001)	(0.002)	(0.002)
	87,707	86,173	42,470	42,161

NOTES: Cell entries are ordinary least squares regression coefficients
for a linear probability model in which the dependent variable is 1 for
reenlistment (additional obligated service of 24 months or more) or 0.
Under the coefficient is its standard error and the number of observations
in the regression. Explanatory variables include the average SRB multiplier
based on cells defined by month of decision, service duty MOS, zone,
and pay grade and, in some specifications, deployment at the time of the
decision. The models include variables for nonhostile deployment, hostile
deployment, service duty MOS fixed effects, years of service at the time
of the decision, educational attainment, gender, AFQT category, race,
an indicator for being promoted more rapidly than is typical, and year-
of-decision indicators (year fixed effects). Robust standard errors are in
parentheses.
** Denotes statistical significance at the 1 percent level.

The Navy, Marine Corps, and Air Force bonus estimates are nearly the same regardless
of whether the SRBM is conditioned on deployment. These services probably did not offer
different bonuses that depended on deployment status. But, as expected, the Army estimates
are much larger when the bonus is conditioned on deployment. The first-term estimate conditioned on deployment is 0.089, over three times larger than when it is not, 0.025.

These estimates bracket the Zone A bonus estimates from both the annual and the
monthly models in Chapter Seven (Tables 7.6 and 7.7); these estimates are 0.059 and 0.039
depending on whether the SRBM is conditioned on deployment. The bracketing seems reasonable. The results in Table 8.1 base the SRBM on whether the individual was deployed in the
month of the reenlistment decision, but in the annual model in Chapter Seven, deployment
refers to being deployed some time within the current fiscal year. A member who reenlists at

the end of the fiscal year may have been deployed earlier in the year but not deployed at the time of the reenlistment decision. Consequently, the upward bias associated with allowing the SRBM to depend on deployment status is attenuated in Chapter Seven to the extent that members who reenlist are not deployed at the time they make their reenlistment decision.[10]

Figures 8.5 and 8.6 depict the estimated SRBM effects on LOR. Figure 8.5 shows estimates not conditioned on deployment, and Figure 8.6 conditions on deployment. The lines in the figures connect the estimated effects of the SRBM on months of reenlistment. That is, each point is an estimated marginal effect of a given bonus level and the lines link them together to show the pattern.[11]

As before, the Army estimates are sensitive to the conditioning but the other services' estimates are not. The Army estimates are much larger when the SRBM is conditioned on deployment, as seen by comparing the Army lines in the figures. The Army estimates in Figure 8.6 are arguably too high, because under stop-loss, the only action that a soldier can take during

Figure 8.5
Effect of an SRBM on Months of Reenlistment, by Service: SRBM Measure Not Conditioned on Deployment, FY 2002 to September 2007

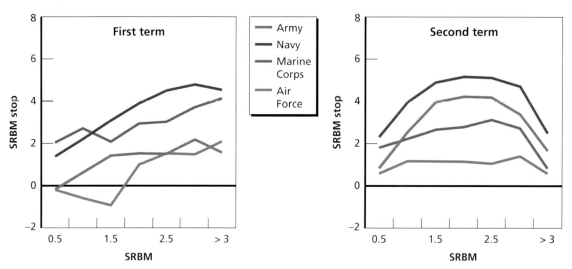

NOTES: Entries are estimated Tobit regression coefficients on the average SRB multiplier based on cells defined by month of decision, service duty MOS, zone and pay grade, and, in Figure 8.3, deployment at the time of the decision. Models control for nonhostile deployment, hostile deployment, service duty MOS fixed effects and years of service at the time of the decision, educational attainment, gender, AFQT category, race, an indicator for being promoted more rapidly than is typical, and year-of-decision indicators. Most effects are statistically significant at the 1 percent level (see Appendix D).
RAND *MG950-8.5*

_____

[10] Another reason the bonus estimates differ from those in Chapter Seven is that the estimates in Table 8.1 use all occupations rather than a subset of occupations. The estimated effect of the SRBM when it is not allowed to vary by deployment for the first term, and using the occupations analyzed in Chapter Seven, is 0.036, which is very close to the Zone A estimate reported in Chapter Seven.

[11] More precisely, these are marginal effects of a given bonus level on the expected value of the LOR conditional on reenlistment, which are computed from the estimates of the Tobit model. The estimated coefficients from the Tobit regression can be found in Tables D.5–D.8.

Figure 8.6
Effect of an SRBM on Length of Reenlistment, by Service: SRBM Measure Conditioned on Deployment, FY 2002 to September 2007

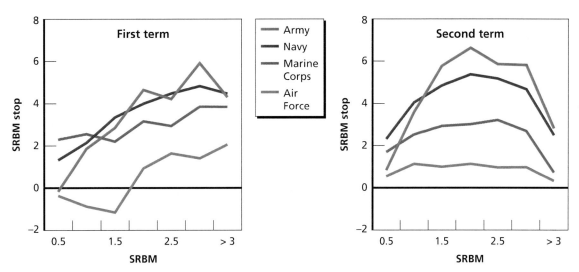

NOTE: See the notes to Figure 8.5.

RAND *MG950-8.6*

deployment is to reenlist, and this is associated with a higher bonus than for soldiers who are not deployed. A main reason for the higher bonus is the deployment bonus, which is available to service members who reenlist when deployed and are in specialties that do not offer an SRB; the bonus for the nondeployed in these specialties is zero. We are not sure whether the Army estimates in Figure 8.5 are too low. They may be biased downward because of reverse causality arising from bonus-setting behavior or measurement error in the bonus variable, or biased upward because of an omitted variable bias coming from the Army's efforts to grow.

Also noteworthy, for all services, the first-term SRBM estimates on months of reenlistment increase as the SRBM increases—a higher bonus multiple leads to an increase in months of reenlistment. At the second term, the estimates for the Army, Navy, and Marine Corps also increase but then decrease, and the Air Force estimates show no increase over the SRBM range from 1 to 3. The downturn in the second-term effects but not the first-term effects is consistent with the idea that each specialty has a maximum allowable bonus, and it is more likely to be reached at the second-term reenlistment, because the pay grade at the second term is higher. If a service member is at or near the maximum, an increase in SRBM could cause a decrease in months of reenlistment. The second-term downturn does not seem consistent with the idea of income effects caused by higher SRBMs. If income effects were the explanation, we might expect to see downturns at the first term as well as the second term.[12] The flatness of the second-term Air Force estimates may reflect an organizational preference to have reenlistments of a fixed length, e.g., four years or six years. In FY 1999, the Air Force chose to limit its

---

[12] Although the differences in the effects between the first and second terms suggest that income effects are not important, strictly speaking, this comparison does not necessarily imply anything about the importance of income effects. One reason is that preferences over the option value of being able to leave the military sooner may differ between the first and second terms. A second reason is that basic pay is lower among first-term personnel, and the additional income provided by a higher SRBM may not be enough to induce them to "purchase" a shorter LOR, whereas it might be large enough for individuals at the second term.

first terms to four or six years, and such a policy at the second term would be consistent. The increase in SRBM might not be enough incentive to shift airmen from four- to six-year commitments. Further, if the minimum LOR is four years, there is no room for an increase from two or three years to four years. Finally, the Army estimates in the figures are similar to those in Chapter Seven.

Estimating the Cost of Additional Reenlistments Generated by Reenlistment Bonuses

We now use the SRBM estimates and bonus information to estimate the cost per additional year of reenlistment generated by bonuses. In the process, we also simulate the effect on reenlistment of decreasing the bonus multiplier to zero. We base the calculations on the bonus level, reenlistment rate, and number of individuals making reenlistment decisions in the first nine months of 2007 (the last calendar year in our data).

As in Chapter Seven, we examine the average cost of an additional year of reenlistment induced by the SRBM using the formula:

$$\text{Average cost of an additional year of reenlistment} =$$
$$\text{Bonuses awarded} / \text{Additional years of reenlistment}$$

where additional years of reenlistment is defined as the change in reenlistment years generated from going from a zero bonus to the observed bonus.

To develop a formula for computation, let r_0 be the observed reenlistment rate in FY 2007 given the average SRBM in FY 2007; r_0 can be thought of as the average probability of reenlistment for an individual at risk of reenlistment. Let L_0 be the average length of reenlistment among those reenlisting in FY 2007, given the SRBM distribution across individuals in FY 2007. The quantity $r_0 L_0$ equals the average unconditional years of reenlistment for a person at risk of reenlisting. Let Δr_0 be the change in reenlistment resulting from a decrease in the bonus multiplier (SRBM) to zero, and let ΔL_0 be the change in LOR resulting from the same decrease. The additional years of reenlistment among those at risk of reenlisting given the SRBM at its FY 2007 level versus its being set to zero is:

$$\text{Additional years of reenlistment} =$$
$$\text{Years of reenlistment in the presence of the bonus} -$$
$$\text{Years of reenlistment in the absence of the bonus} =$$
$$r_0 L_0 - (r_0 - \Delta r)(L_0 - \Delta L)$$

We calculated this quantity using the observed FY 2007 reenlistment rate and average LOR among those reenlisting and the estimated effects reported in Table 8.1 for the reenlistment margin and Appendix D for the LOR margin:

$$\text{Additional years of reenlistment} =$$

$$r_0 L_0 - (r_0 - \beta_r \overline{SRBM}) \left(L_0 - \sum_{j=0.5}^{>3} \beta_{L,j} \theta_{SRBM=j} \right)$$

where β_r is the estimated reenlistment margin effect of a one-unit bonus step increase, $\overline{SRBM}$ is the observed mean SRBM in FY 2007, $\beta_{L,j}$ are the LOR margin effects for SRBM at level j, $j = 0.5, 1.0, \ldots, 3.0, >3$, and $\theta_{SRBM=j}$ is the fraction of those reenlisting who received a SRBM at level j. We computed the bonus distribution for both versions of our SRBM variable (see Appendix D for these tabulations).

Table 8.2 shows the components that enter into the cost calculations. The Army cost estimates are sensitive to whether the SRBM variable used in the regression models varies by deployment status. When the SRBM does not vary by deployment, the expected change in reenlistment is smaller, which leads to higher cost estimates relative to those when the SRBM does vary by deployment. For the Army, the estimated average cost of bonus expenditures per additional year of reenlistment is $24,862 for the first term and $23,900 for the second term. The second-term cost is slightly lower despite the fact that the SRBM effects on first- and second-term Army reenlistment are identical. The lower cost occurs because the second-term SRBM effect on LOR is larger than the first-term effect (see Figure 8.5), and this turns out to be enough to offset the higher bonus cost at the second term deriving from higher average pay grade. If the larger SRBM coefficients are used, the average costs are $8,282 and $15,513, respectively. (Below we discuss these Army cost estimates and compare them with those in Chapter Seven.)

The estimated costs per additional year of reenlistment at the first term for the other services are $13,907 to $17,033 for the Marine Corps, $24,737 to $28,007 for the Navy, and $67,378 to $70,242 for the Air Force. The Marine Corps estimates are within the range of the Army estimates, but the Air Force estimates are much higher. This is because the SRBM effects on reenlistment and LOR are smaller for the Air Force than for the other services. The Navy cost is somewhat higher than for the Army or Marine Corps although this difference is not large.

Table 8.2
Calculation of the Average Cost per Additional Year of Reenlistment Generated by a Bonus, FY 2007

	First Term				Second Term			
	Army	Navy	Marine Corps	Air Force	Army	Navy	Marine Corps	Air Force
Average SRBM in FY 2007 (including 0)	1.50	1.17	2.69	0.61	1.70	1.91	1.34	1.33
FY 2007 reenlistment rate	0.39	0.43	0.37	0.61	0.81	0.70	0.78	0.78
Average LOR (years) among reenlisters	3.43	3.42	3.86	4.09	3.64	3.85	3.65	4.21
Average bonus per decision	$4,515	$4,652	$7,627	$2,794	$12,142	$14,418	$9,591	$14,742
Change in Unconditional Mean LOR (Years) Generated by FY 2007 Bonus								
SRBM does not vary by deployment	0.18	0.17	0.45	0.04	0.51	0.37	0.12	0.14
SRBM varies by deployment	0.55	0.19	0.55	0.04	0.78	0.38	0.13	0.13
Average Cost per Additional Year of Service Generated by FY 2007 Bonus								
SRBM does not vary by deployment	$24,862	$28,007	$17,033	$70,242	$23,900	$38,897	$77,455	$101,880
SRBM varies by deployment	$8,282	$24,737	$13,907	$67,378	$15,513	$38,139	$74,649	$112,275

The average costs are also higher for the second term than for the first term. The second-term estimates are about $39,000 for the Navy and range from $102,000 to $112,000 for the Air Force. The largest difference relative to first term occurs in the Marine Corps. Its second-term cost estimate of about $75,000 is several times larger than the first term cost estimate. This occurs because the SRBM effect on second-term reenlistment is virtually zero and only the increase in LOR with SRBM works to prevent the cost from being much higher. The Air Force cost is high because of a smaller SRBM effect on reenlistment (than in the Army and Navy) and the lowest SRBM effects on LOR.

How do our Army estimates compare with those in Chapter Seven? The cost estimates in Chapter Seven are $8,678 for Zone A and $10,409 for Zone B, where the SRB multiplier varies by deployment, and between $11,870 and $15,729 for Zone A and between $13,383 and $20,075 for Zone B, where the SRB multiplier does not vary by deployment. The latter range is determined by different assumptions about the effects of multiplier changes on LOR. These Zone A and Zone B estimates are similar to the intervals defined by the estimates reported in Table 8.2.

When we adjust for differences in the construction of the estimates between this chapter and Chapter Seven, the estimates become closer, although several differences remain. First, the estimation methodologies are different; Chapter Seven uses a hazard model to estimate the reenlistment margin response, whereas a linear probability model is used here. Thus, the estimated effects of the SRBM differ. Second, we use the FY 2007 reenlistment rate, LOR, and bonuses, whereas Chapter Seven uses FY 2006 values. If we use the SRBM effects from Chapter Seven and apply the FY 2007 data to them, the cost estimates are $9,824 for the first term and $13,246 for the second term. Third, the estimates in Table 8.2 are based on all occupations rather than on a subset of occupations. We reestimated our models on the 24 occupations used in Chapter Seven and made new cost calculations. We found that the cost implied by the coefficients in the models where the SRBM varies by deployment was $5,941 for the first term and $12,531 for the second term, and for models where the SRBM does not vary by deployment, the cost was $18,740 for the first term and $20,765 for the second term. In short, the Zone A and Zone B cost estimates from Chapter Seven are very close to those we compute for the Army using estimates reported in this chapter.

Concluding Remarks

In this chapter, we discussed a reenlistment model estimated for all services that relates the likelihood of reenlisting to deployment experiences, demographic characteristics, and the bonus level available to service members at the time they make their reenlistment decision. This research extends the work in HM, notably by using refined bonus measures that more accurately reflect bonus generosity. We also estimate bonus effects on the LOR conditional on reenlisting. These refinements and models make the analysis in this chapter very similar to that in Chapter Seven even though the analyses originated independently.

We estimate positive effects of the SRBM on reenlistment in all cases except for second-term marines, where the effect is statistically zero. However, there is variability in the magnitude of these estimates. For the Army, we estimate that a one-unit change in the SRBM changes the probability of enlisting by 0.025 percentage points, which is in line with earlier estimates. When the SRBM varies by deployment, the estimated effects on reenlistment and on LOR increase sharply for the Army, yet we think these estimates are biased upward (and the cost estimates biased downward) for reasons related to stop-loss and the payment of deployment bonuses. For the other services, the choice of whether the SRBM variable varies by deployment has relatively little effect on the estimated effects. Further, the estimates on LOR indicate that increases in SRBM increase months of reenlistment. This positive relationship is quite clear at first-term reenlistment, but at second-term reenlistment, the relationship turns down at high levels of the bonus variable. We think that the reason for the downturn lies in bonus ceilings that are more likely to limit bonus payments at the second term than at the first term, which is consistent with findings in Chapter Seven. The Air Force effects differ from this pattern, however. There is no increase in LOR as the bonus multiple increases from one to three. This lack of responsiveness to SRBM might result from a policy constraining LOR to be four or six years, with the SRBM increase not providing enough incentive to shift many airmen from four to six years, but this is speculation and requires further research.[13]

Since FY 2001, bonuses have increased sharply for the Marine Corps and Army. In this chapter, we calculate the number of additional years of reenlistment generated by these increases, and estimate the average cost of an additional year of reenlistment. For the Army, the estimated bonus effects depend on whether the SRBM variable is conditioned on deployment. We suggest that conditioning on deployment likely results in upward-biased estimates. For the other services, the estimates differ little by whether the SRBM variable is conditioned on deployment.

Given the likelihood that bonus ceilings cause a downturn in the effect of SRBM on LOR at high levels of SRBM, there is an implication for policy. This is to raise the bonus ceilings, especially at the second term. This can be done selectively and should be done in occupations where the services want to achieve longer service commitments. Longer commitments would decrease the cost per additional year of reenlistment.

---

[13] In our data, 45 percent of first-term and 60 percent of second-term Air Force reenlistments are for four years, 26 percent of first-term and 23 percent of second-term reenlistments are for five years, and 12 percent of first-term and 5 percent of second-term reenlistments are for six years (note that, in our data, individuals sometimes reenlist a month or two before their ETS date, so that a four-year reenlistment appears as a net increase in 46 or 47 months to ETS, and similarly for five- and six-year reenlistments). In comparison, there is less "clumping" of the LOR at four and six years in the other services, particularly the Army. This is consistent with the conjecture that the effects on LOR for the Air Force are smaller, because choices over LOR are more constrained in the Air Force.

CHAPTER NINE

Conclusions

The General Accountability Office recently assessed the Army's use of cash incentives, specifically enlistment and reenlistment bonuses, and concluded that the Army does not know if it is paying more than it needs to pay in bonuses and, therefore, if these programs are as cost-effective as they can be. As part of the 2009 Defense Appropriations Act, Congress expressed concern that the size and scope of cash incentives have increased without performance metrics to determine whether they are cost-effective. Congress therefore requested that DoD report on the number and amount of these cash incentives, as well as on metrics of performance. This document provides input to the DoD report.

The purpose of the bonus programs is to help manage the enlistment and reenlistment of trained and experienced personnel. We develop metrics of performance based on the results of deliberations of past commissions and study groups on military compensation. The 7th Quadrennial Review of Military Compensation articulated the objectives of the military compensation system and the 2006 Report of the Defense Advisory Committee on Military Compensation built on these objectives to state a set of principles for evaluating and guiding change to the military compensation system (DoD, 2006, pp. 9–12). These principles include the requirements that the compensation system be

1. linked to force management objectives, particularly recruiting and retention objectives
2. flexible, by adjusting quickly to circumstances affecting the supply and demand for personnel and by addressing specific problems in specific areas
3. market-based and consistent with the idea of choice and volunteerisms, that is, the compensation system should embed incentives for members to volunteer for assignments, deployments, occupations, and other aspects of military service
4. efficient, by meeting force management objectives in the least costly manner.

The rest of this chapter assesses the performance of the EB and SRB programs in terms of these principles, drawing from the results in previous chapters. We first discuss our estimates of the extent to which these enlistment and reenlistment bonuses contributed to the ability of the Army to meet its recruiting and retention objectives (principle 1). We then discuss the degree to which these incentives were used in a flexible manner, in terms of adjusting quickly to changing market conditions, as well as specific problem areas (principle 2). Inherent in both of these discussions is the idea that EBs and SRBs are incentives that guide the free choice of personnel to volunteer for service and to select specific occupations (principle 3). Finally, we discuss our results pertaining to the cost-effectiveness of these incentives (principle 4), specifically, whether economic rents were paid, i.e., whether bonuses were paid to individuals who would have enlisted or reenlisted at lower bonus levels.

Caveats

Before providing our assessment, we note that our analysis does not provide estimates of the absolute cost of bonuses or whether bonuses were optimal in the sense that the same levels and mix of enlistments and reenlistments across different occupations and enlistment or reenlistment terms could be produced at lower cost if bonuses were managed differently. As discussed below in the subsection on future research, the optimal levels could be evaluated, but this is best done within the context of an experiment. Such an analysis is beyond the scope of this research.

We also reiterate a point made throughout the report with respect to our estimates, namely, that they are subject to a number of biases in the sense that the estimated effects of bonuses may deviate from the true effects of bonuses, either in an upward direction (thereby overstating the effects of bonuses) or in a downward direction (thereby understating the effects of bonuses). Such biases may stem from a number of sources. The first source of potential bias results from the possibility of reverse causality, a term used in the econometrics literature to refer to cases where a variable not only affects the outcomes of interest, but the outcomes of interest also affect the variable. In our case, the variable is bonuses and the outcomes of interest are enlistments and reenlistments. Bonuses influence the willingness to enlist or reenlist, but enlistment and reenlistment outcomes also may influence the amount of bonuses set by policymakers. This phenomenon imparts a downward bias to our estimates. Another source of potential bias is that additional factors that are not observed in our data and that are correlated with bonuses may increase enlistments and reenlistments. Omitting these other factors can impart an upward bias on the estimated bonus effects, offsetting to some extent the potential downward bias associated with reverse causality.

Yet another potential source of bias arises from the definition of the SRB multipliers and whether it depends on deployment status. The estimated effect of the SRB multiplier is biased upward if it depends on deployment status and if higher bonuses are associated with a greater chance of reenlistment as a result of deployment status and deployed personnel are more likely to reenlist. On the other hand, the estimated effect of the SRB multiplier is biased downward if it does not depend on deployment status, because of measurement error.

We attempt to attenuate some of these biases in our estimating methodology, or we attempt to establish an upper and lower bound by showing a range of estimates. Nonetheless, there is uncertainty about the true estimates, and the estimates presented in this report must be interpreted as associations between bonuses and enlistment or reenlistment. That said, the estimates are quite robust and are quite consistent with past estimates found in the literature.

Furthermore, we note that using administrative data to estimate the effects of bonuses has several advantages. For example, such an analysis can be completed relatively quickly, given the ease of access to such data, and it permits analysis of other variables of interest, such the effects of recruiting resources, civilian employment opportunities, and the Iraq War, in the case of enlistment, and the effects of deployment, in the case of reenlistment.

An alternative approach would be to conduct an experiment, similar to the enlistment bonus test conducted in the early 1980s. Experiments have numerous advantages, including the ability to eliminate the bias created by reverse causality, because personnel can be randomly selected to receive different bonus amounts. On the other hand, as discussed by Moffitt (2004) and Heckman and Smith (1995), experiments are not a perfect solution, because they also have drawbacks. For example, estimates produced from experiments may be biased by contamination of the control group or if the treatment is implemented differently in different sites. In addition, experiments may provide little information about other factors affecting outcomes or little information about the mechanism that leads to the effects that are estimated, unless the experiment is specifically designed to analyze these other factors or the mechanism underlying the estimated effects. In other words, experiments can be a "black box." Because of the advantages and disadvantages of the nonexperimental approach used here as well as experimental approaches, and because problems with each approach are addressed to some extent by the other approach, ideally both approaches should be used.

Did Bonuses Enable the Services to Meet Their Recruiting and Retention Objectives?

Because the GAO report focused on the Army and because the Army accounts for much of DoD's bonus expenditures, we focus first on the Army. We used our estimated Army enlistment model to simulate what would have happened to Army high-quality enlistments between October 2004 and September 2008 in the absence of the increase in the average enlistment bonus over this period. Among those receiving enlistment bonuses, the average bonus increased from about $5,600 to about $18,000 over this period. The share of Army enlistments receiving bonuses increased from about 50 percent to 70 percent.

We project that over this four-year period, the bonus expansion increased high-quality enlistments by 26,700, or 20 percent of the actual number of high-quality enlistments achieved by the Army. That is, in the absence of the increase in bonuses over this period, the Army would have fallen short by 20 percent and would not have been able to meet its recruiting goal. This simulation focuses on the market expansion effects of bonuses and does not include the effects of enlistment bonuses on channeling recruits into critical occupational areas.

We also examined the enlistment effects of the Navy bonus program and projected the pattern of high-quality Navy enlistments in the absence of increases in bonuses between FY 2004 and FY 2008. We find that bonuses had a minimal market expansion effect—an increase of about 1,700 enlistments over the period.

Our estimated models for each service also allow us to simulate the effects of setting the reenlistment bonus multiplier to zero in FY 2007, a move that would have eliminated the SRB program in that year. In FY 2007, the average unconditional first-term bonus multiplier was 1.5 for the Army and 2.69 for the Marine Corps. For the Army, our estimates differ depending on whether the SRBM varies with deployment. We project that eliminating the SRB program in that year for the Army would have reduced the probability of reenlistment from 39 percent to 35.3 percent using our estimates that do not depend on deployment status, and to 25.7 percent using our estimates that do depend on deployment status. For the other services, the estimates do not differ when SRBM depends on deployment. Eliminating the Marine Corps SRB program in FY 2007 would have reduced the first-term reenlistment rate from 37 percent to 27 percent. The simulated effects for the Air Force and Navy are more modest because the average bonus multipliers were lower and the average reenlistment rates were higher.

The simulated drops associated with eliminating the bonus program in the Army and Marine Corps in first-term reenlistment rates are large. We find a similar result when we use the Army model estimates for the 24 MOS. Thus, we conclude that in the absence of the reenlistment bonus program, reenlistments would have been much lower. As these branches had the heaviest combat duties in OEF and OIF, bonuses helped compensate for the heavy deployments. Together with our results on the effects of the Army enlistment bonus program, we conclude that the enlistment bonus and reenlistment bonus programs contributed significantly to the Army's recruiting and retention success in recent years.

Estimates for the Navy, Air Force, and Marine Corps at the second reenlistment point suggest that bonuses had a much more modest effect on reenlistments. The Air Force generally experiences the smallest bonus effects, although the estimated bonus effects for the Marine Corps are negative but not statistically different from zero for the second term. In general, the estimated effects of bonuses on second-term reenlistment are smaller than they are on first-term reenlistment. This may reflect the fact that significant self-selection on taste for service has already occurred at the end of the first term, so reenlistment rates at the end of the second term are less responsive to financial incentives. Furthermore, although the estimated effect is lower, in percentage terms, the reenlistment rates are higher at the end of the second term. Thus, a smaller effect, in percentage terms, can result in the same number of reenlistments, in absolute terms, at the end of the second term as at the end of the first term.

We also find that bonuses increase the length of reenlistment chosen by reenlistees, but the effect diminishes at higher SRB multipliers, particularly at the second term (Zone B). The bonus effect on LOR is smallest for the Air Force and the Marine Corps at the end of the second term. The diminishing effect of bonuses on LOR means that as the SRB multiplier increases, the length of obligated service increases, but at higher multiplier levels, the positive effect on the length of obligation declines. We find this for all services at the second reenlistment point in the models estimated that are based on HM and for both the first and second reenlistment point in the Army model based on the 24-MOS sample.

The decreasing estimated effect of reenlistment bonuses on length of obligation at higher bonus multipliers may be explained by several factors. First, the services place caps on bonus amounts so that, at some point, members have no incentive to choose longer terms, since doing so has no effect on the bonus amount. We find evidence to support this explanation for the Army. When the SRB caps were higher before FY 2005, the diminishing effect of the SRB multiplier on length of service is found to occur at higher multiplier levels, just as we would expect if the caps were less constraining. Second, service policy may limit the ability of members to choose term length, especially in some occupational areas. A service might expect or constrain the service member to choose a four- or six-year reenlistment, and increases in the bonus multiple might have little effect on the length chosen. This may be the case for the Air Force, which has the smallest bonus effect on length of reenlistment. Third, bonuses may have a diminishing effect on LOR as the multiplier increases, because reenlistees faced with a higher multiplier may choose shorter term lengths that give them the flexibility to leave earlier to take advantage of civilian opportunities.

The first two of these explanations suggest the possibility of improving the effectiveness of reenlistment bonuses. Bonus caps could be more actively managed so that increases in multipliers do not provide incentives to choose term lengths that are shorter. In addition, the services could change their policies to allow members to have more flexibility to increase term length as the multiplier increases.

We note that our results on the effects of SRB multipliers on length of reenlistment are estimates of the effects on number of obligated years and not necessarily on length of service. Thus, we find that higher SRB multipliers increase the number of years reenlistees choose to obligate, although the effect diminishes in the second term. This does not mean that the number of years they ultimately stay in service is shorter at higher multiplier levels. At higher multiplier levels, individuals may choose shorter obligations but may ultimately serve the same number of years in their career as those choosing longer obligations. The former may simply choose more reenlistment terms of shorter length whereas the latter may choose fewer terms of longer length.

Were Bonuses Used in a Flexible Manner?

As discussed in the reports of the 9th and 10th Quadrennial Review of Military Compensation, the key role of bonuses, and special and incentive pays more generally, is to add flexibility to the military compensation system. These pays are intended to selectively address specific force management needs, such as staffing shortfalls in particular occupational areas, hazardous or otherwise less-desirable duty assignments, and attainment and retention of valuable skills. They are also intended to allow the compensation system to adjust quickly and temporarily to supply and demand changes in personnel.

We assess whether enlistment and reenlistment are used flexibly by considering the degree to which the Army (in the case of enlistment and reenlistment bonuses) and the Navy (in the case of enlistment bonuses) have targeted these incentives to specific groups, i.e., have bonuses served a skill-channeling purpose and how quickly have they changed?

Consider first enlistment bonuses. The share of Army enlistments receiving bonuses rose from about 40 percent in September 2004 to about 70 percent in September 2008. Thus, in the case of the Army, enlistment bonuses increasingly were used to expand the market. That said, the dollar value of enlistment bonuses among those receiving them varied across occupations, even in FY 2008 when most enlistees received bonuses. As shown in Figure 2.4 for a selected set of occupations and as listed in Table A.1 for all Army occupations, average bonuses increased between FY 2004 and FY 2008 for occupations that have traditionally received bonuses, such as combat arms. Furthermore, occupations that had little history of bonuses also began to receive them, such as law enforcement and religious services.

Nonetheless, critical occupational specialties, such as infantry, field artillery, and air defense, received substantially larger bonuses than other occupational areas. For example, in FY 2008, fire support specialists (13F) received an average bonus of $18,700, psychological operations specialists (37F) received an average bonus of $17,500, and explosive ordnance disposal specialists (89B) received an average bonus of $19,200. These were among the highest average bonuses offered in FY 2008. At the low end, carpentry and masonry specialists received an average bonus of $3,500, armament repairers received a bonus of $2,800 on average, and cargo specialists received an average bonus of $2,500.

Substantial differences in bonuses also were evident across terms of service. Enlistees who contracted for a term of six years of service received an average bonus of $15,100 in FY 2008, those who contracted for five years received an average bonus of $11,000, and those who contracted for four years received an average bonus of $8,700. Thus, the premium for a six-year enlistment (relative to a five-year enlistment) was about $4,200 and for a five-year enlistment (relative to a four-year enlistment) was about $2,300. Furthermore, at the same time that most Army enlistees received bonuses in FY 2008, the premium for choosing a longer enlistment term increased. The premium for an additional year of enlistment in FY 2006 was $4,500 (in FY 2008 dollars).

These results suggest that at the same time the Army used bonuses to expand the market, it also used them to differentiate occupations and to provide incentives to enlistees to choose specific occupations and longer enlistment lengths. We do not assess the effects of enlistment bonuses on the probability of choosing any specific occupation, but analysis of the skill-channeling effects of bonuses from the 1980s indicates that these effects are large (Polich, Dertouzos, and Press, 1986).

In contrast to the Army, the Navy expanded average bonuses only modestly between FY 2004 and FY 2008, with the share of enlistees receiving bonuses actually declining over this period. Thus, bonuses played less of a market expansion role for the Navy than the Army over this period. On the other hand, average Navy bonuses differed substantially across occupational areas. As discussed in Chapter Two and shown in Table A.3, average bonuses for seamen temporarily increased in FY 2006, average bonuses for cryptologic technicians grew steadily between FY 1999 and FY 2008, and bonuses for other ratings declined.

Now consider reenlistment. It is clear that the Army used reenlistment bonuses for a period of time as an across-the-board tool to ensure that it met its retention targets, at least in FY 2007, when 80 percent of reenlistees received an SRB, and even early in FY 2008, when eligibility for the enhanced SRB was predicted to be 81 percent. On the other hand, as is apparent from the discussion in Chapter Six, the Army also used complex rules to fine-tune the targeting of the dollar amount of SRBs to specific groups. The amounts of the SRBs were targeted based on the occupation, rank, length of reenlistment of reenlistees, as well as their skill (within an occupation), location, unit assignment, and deployment status. Clearly, the Army used these incentives in a flexible manner to induce the reenlistment of trained and experienced personnel in different occupations, skill areas, units, and combat zones.

Finally, proper EB and SRB management requires not only ramping them up when recruiting and retention deteriorate but turning them off when recruiting and retention improve. That is, these incentives should be adjusted quickly as circumstances change. The number of SRB program changes that the Army announced each year in the FY 2001–2008 period is evidence that the Army seemed to manage its SRB program proactively. Furthermore, the SRB reductions in the FY 2002–2003 time frame and the substantial reductions announced in March 2008 show that the Army is not reluctant to reduce SRBs when retention is above target.

Bonus variation over time is also evident in the other services. The Air Force substantially reduced the percentage of reenlistees who received a bonus at the end of the first term between FY 2002 and FY 2007 but substantially increased the average bonus multiplier for those who did receive a bonus. In contrast, the Navy reduced the average bonus multiplier and the percentage receiving bonuses for those at the end of the first term between FY 2000 and FY 2008. As with the Army, the Marine Corps increased both the percentage of reenlistees receiving a first-term bonus and the average bonus multiplier. These service differences show that the services use bonuses to respond to service-specific conditions and obviate the need for across-the-board responses.

We conclude this subsection by noting that most elements of compensation are common to the four military services. But, as the operations in Iraq and Afghanistan reveal, each service may from time to time be confronted with service-specific shocks to recruiting and retention.

Indeed, Army recruiting and retention have been the most affected by the operations in these countries. When managed properly, the enlistment and reenlistment bonus programs of the four services can act as residual adjustment mechanisms and obviate the need for compensation adjustments that are not service-specific. That a high percentage of enlistees and reenlistees receive a bonus need not be viewed as an unnecessary or excessive use of bonuses if in fact their high use is in response to an overall, service-level shortfall of enlistments and reenlistments.[1]

Were Bonuses Used Cost-Effectively?

We measure the cost-effectiveness of bonuses by comparing the cost of additional recruits or reenlistees receiving bonuses to generate the added supply of personnel to the cost of other resources. Because we judge cost-effectiveness using a relative and not an absolute measure, our measure provides no information about whether bonuses were set too high for too long. That is, we do not assess whether bonus levels were optimal or at their most efficient levels. Instead, we assess whether the additional personnel that the services recruited and retained could have been recruited and retained at less cost using pay, or with alternative recruiting resources, instead of bonuses.

In the case of enlistment in the Army, we estimate that EBs are more cost-effective than pay but less cost-effective than recruiters as a way to expand the market. We estimate a marginal cost of enlistment bonuses of $44,900 and of $57,600 for pay. We also estimate a lower marginal cost for Army recruiters, of $33,200. Thus, we find that bonuses are relatively more cost-effective than pay but less cost-effective than recruiters. However, for several reasons we believe that we overstate the marginal cost of EBs. First, we include only the market expansion effects of bonuses and not their skill-channeling effect. Second, we estimate that enlistment bonuses have a small but statistically significant effect on reducing attrition, thereby increasing person-years, but we do not incorporate this increase in person-years when we estimate marginal cost. Third, we do not account for the effects of enlistment bonuses on enlistment term length. If bonuses increase term length, our estimates overstate the total marginal cost of bonuses. Similarly, our estimate of the cost-effectiveness of recruiters does not incorporate the benefits associated with the ability of the services to flexibly target them at different geographic regions, unlike bonuses or pay, or the disadvantages of recruiters that arise because the size of the recruiter force cannot be changed quickly because of the time it takes to train new recruiters.

Relative to pay and recruiters, bonuses can be more directly targeted at particular occupations, and bonuses can also be changed relatively quickly to respond to short-run developments in the recruiting environment. However, bonuses are potentially more likely than other incentives to generate "skimming effects," whereby bonuses offered in one service attract recruits who may have otherwise joined a different service. Therefore, although cost-effectiveness is one criterion for comparing recruiting resources, other considerations may also be important.

---

[1] On this point, it is useful to note that the British now have separate basic pay tables for their military services, with higher basic pay for the Royal Army and Royal Marines than for the Royal Navy and Royal Air Force. There would no doubt be much resistance to separate pay tables for the U.S. military services. But such differences are not needed as long as there is a residual element of compensation that can be adjusted flexibly and as needed. In the United States, bonuses can serve as a service-wide adjustment mechanism and indeed appear to have done so recently.

In the case of reenlistment, we provide a range of estimates of the marginal cost of SRBs using alternative assumptions and using different SRB estimates, depending on whether the SRB multiplier includes deployment. The estimates account for both the effects of SRB multipliers on reenlistment and length of reenlistment. For the Army, our estimate of the marginal cost of a change in the SRB multiplier at the first reenlistment point is in the range of $8,300 to $24,900 per person-year. Our estimates are in the range of $10,400 to $23,900 per person-year for the second term.

The range of estimates reflects our different approaches and data sets as well as whether the SRBM is defined as conditional on deployment status and our choice of the estimated effect of the SRB multiplier on length of reenlistment. The lower and upper bounds are probably unrealistically low and unrealistically high, given our concern about potential biases in our SRBM estimates. Thus, we believe that within the $8,300 to $24,900 range, the most reasonable marginal cost estimate for the Army at the first term is between $8,300 and $11,900. The $8,300 figure is our Army estimate based on the HM model and where the SRB multiplier is not conditional on deployment status, and the $11,900 figure is our estimate based on the Army 24-MOS sample where the SRB multiplier is conditional on deployment. Similarly, we believe that the most reasonable range for the marginal cost of reenlistment bonuses for the Army at the second term is between $13,400 and $15,500.

The estimates of the change in cost per additional person-year at the first term for the other services are $13,900 to $17,000 for the Marine Corps, $24,700 to $28,000 for the Navy, and $67,400 to $70,200 for the Air Force. Thus, the Marine Corps cost is roughly similar to the Army cost, and the Air Force cost is substantially higher. The Navy cost is somewhat higher than that for the Army or Marine Corps, although the difference is not large.

The marginal cost estimates per person-year are also higher for the second term than for the first term, as in the Army. The highest estimates are for the Air Force, where the estimates range from $101,900 to $112,300, and for the Marine Corps, where the estimate is about $75,000. The higher marginal cost estimates for these services reflect lower responsiveness to SRB multipliers of reenlistments and length of reenlistment. Furthermore, reenlistment rates are higher for the Navy and Air Force, so costs are higher, reflecting more rents paid, when bonuses are increased.

For the Army, Navy, and Marine Corps at the end of the first term, we find that reenlistment rates and length of reenlistment are responsive to SRBs, and the responsiveness is sufficient to make SRBs a cost-effective policy instrument relative to pay despite the fact that reenlistees receive economic rents (payments in excess of opportunity costs) when SRBs are increased. The reason is that bonuses are targeted at subgroups of personnel and can be increased and lowered quickly as needed, so the amount of economic rent required to induce a given number of enlistments or reenlistments is lower when bonuses rather than pay are used to generate the increase in supply at the end of the first term for these services.

Similarly, we find that SRBs are a cost-effective policy instrument relative to pay at the end of the second term for the Army and Navy. For the Air Force, for both the end of the first and second terms, and for the Marine Corps, at the end of the second term, the cost estimates are quite high. Taken literally, these estimates indicate that bonuses are a costly way of obtaining additional person-years for these services at these reenlistment points. However, we urge caution in drawing this conclusion for two reasons. First, our estimated bonus effects may be biased downward, subject to bonus caps, and may reflect limited flexibility to choose term length, as discussed above. A downward bias in the estimated bonus effects means that the estimates are too small, so estimates of marginal cost are too high. Second, the cost-effectiveness of bonuses can be deemed too high only relative to a benchmark. Our benchmark is the cost of achieving a given reenlistment target using an alternative approach, namely, with an across-the-board pay raise. An across-the-board pay increase applies to all occupations, not just those with an impending shortage; creates a higher pay floor, which might mean higher pay costs in all future years; and gives the same pay increase to everyone. Military pay must be kept competitive overall, and pay increases provide the foundation for competitiveness. Bonuses allow for selective increases to differentiate pay by occupation and experience level and can be easily increased or decreased depending on current conditions. The estimated bonus costs are likely to be substantially less than the marginal cost of raising military pay to achieve reenlistment goals.

Is There Room for Improvement?

Our main conclusion is that the Army used enlistment and reenlistment bonuses effectively to meet its recruiting and retention objectives, managed the programs flexibly in terms of targeting them at specific groups, and adjusted them in a timely manner as recruiting and retention changed. We also conclude that for the Army, these programs are relatively more cost-effective than other recruiting and retention resources. For the other services, we conclude that they also used reenlistment bonuses in a flexible manner by varying bonuses over time and that bonuses are also relatively more cost-effective than pay in achieving reenlistment goals. That said, our analysis does not provide information on whether bonuses were optimally set.

Even so, our analysis does suggest an area where the management of these programs could be improved. Our analysis provides evidence that at high levels of SRB multipliers, reenlistees choose shorter term lengths. Although this may result for a number of reasons, two possibilities suggest that the services could improve their management of bonus caps and provide members with an incentive to choose longer terms and the flexibility to do so.

Areas for Future Research

A key remaining question is whether differentials in bonuses across occupations are too large and whether bonuses are larger than needed to sort personnel into critical occupations. That is, although bonuses are generally cost-effective in improving enlistments and reenlistments relative to pay, more information is needed about whether the services are implementing bonuses in the most efficient manner. Additional analysis is needed on whether a different mix and different levels of bonuses than those actually observed would have resulted in more enlistments and reenlistments for the same cost. Such an analysis would require occupation-specific estimates of the effects of enlistment and reenlistment bonuses. Such estimates are best obtained in an experimental setting where many of the potential biases discussed earlier could be eliminated or attenuated. Such an analysis was beyond the scope of the current study and should be pursued in future research. Our analysis suggests that the management of bonus caps might be improved. However, additional analysis is needed to understand the implications for cost-effectiveness of the bonus caps.

In the case of enlistment, more information is needed about the skill-channeling effects of bonuses. In the case of reenlistment, information is needed on the relative productivities of personnel in different occupations and on how productivity grows with experience in different occupations. Such information is necessary to assess the payoff, in terms of productive person-years, of bonus differentials across different occupations as well as in lengths of reenlistment. In addition, for both enlistment and reenlistment, more analysis is needed to understand why the responsiveness to bonuses found in this report differs across services.

We also recommend that Congress and DoD conduct an experiment to determine the effectiveness of bonuses. Such a test would supplement the estimates in this report of the effects of bonuses on enlistment and reenlistment and could provide information on the effects of bonuses without the confounding effects of different sources of bias discussed above. Furthermore, an experimental approach would permit estimates of the skill-channeling effects of bonuses on enlistment. In the 1980s, DoD conducted an enlistment bonus experiment and that research provided a strong foundation for bonus policy for at least two decades. To answer queries by Congress about bonus effectiveness in the future, a new bonus test should be conducted to supplement the analysis of administrative data presented in this report.

Detailed Background on Enlistment Bonuses

Table A.1
Average Bonuses for Army Enlistees, by Occupation, FY 2000–2008

Code	Description	2000	2001	2002	2003	2004	2005	2006	2007	2008
09S	Commissioned Officer Candidate	N/A	N/A	0	82	335	4,060	2,176	0	0
09W	Warrant Officer Candidate	N/A	N/A	N/A	N/A	0	0	0	0	189
11X	Infantry Recruit	2,597	5,492	5,213	2,755	3,655	9,838	10,453	9,865	14,348
13B	Cannon Crewmember	2,877	4,736	5,532	3,916	4,982	9,003	11,428	12,328	16,750
13C	Automated Fire Support Systems Specialist	302	1,844	N/A	N/A	N/A	N/A	N/A	N/A	N/A
13D	Field Artillery Tactical Data System Specialist	1,261	2,870	2,368	1,843	2,767	8,479	14,902	12,929	24,170
13E	Cannon Fire Direction Specialist	450	1,741	N/A	N/A	N/A	N/A	N/A	N/A	N/A
13F	Fire Support Specialist	8,545	6,900	4,105	3,249	6,021	11,394	13,512	18,681	23,961
13M	Multiple Launch Rocket System (MLRS) Crewmember	8,557	7,757	3,886	4,492	6,352	9,659	9,705	7,406	9,167
13P	Multiple Launch Rocket System Operator/Fire Direction Specialist	2,876	5,655	3,766	2,065	2,048	4,076	6,410	3,326	6,216
13R	Field Artillery Firefinder Radar Operator	1,525	3,163	2,149	1,025	1,948	7,184	14,496	12,997	26,718
13S	Field Artillery Surveyor	463	1,335	1,171	1,069	639	4,615	4,077	N/A	N/A
13W	Field Artillery Meteorological Crewmember	265	242	0	91	250	3,367	4,840	6,019	9,487
13X	Field Artillery Computer Systems Specialist	N/A	2,041	2,326	N/A	N/A	N/A	N/A	N/A	N/A
14J	Air Defense Tactical Operator	1,701	2,646	2,879	2,715	3,218	11,383	15,142	16,659	26,401
14R	Bradley Linebacker Crewmember	802	5,942	7,230	2,812	3,272	N/A	N/A	N/A	N/A
14S	Air and Missile Defender	765	1,229	969	528	664	1,536	2,468	7,012	5,965
14T	Patriot Launching Station Enhanced Operator/Maintainer	3,361	3,950	3,874	2,568	5,075	5,938,	12,529	14489,	23,360
14Z	Air Defense Artillery Senior Sergeant	5,035	6,674	5,107	3,506	4,068	6,959	13,639	15,144	28,368
15B	Aircraft Powerplant Repairer	636	902	872	147	282	2,929	2,866	5,704	6,667
15D	Aircraft Powertrain Repairer	885	742	758	672	987	2,160	2,179	3,652	4,412
15F	Aircraft Electrician	763	759	179	0	112	2,524	2,622	4,300	1,536
15G	Aircraft Structural Repairer	1,732	4,151	3,735	1,708	1,048	2,919	2,313	5,885	6,374
15H	Aircraft Pneudraulics	1,428	1,057	1,493	810	410	1,113	1,287	3,691	1,268
15J	Oh–58D Armament/Electrical/ Avionics Systems Repairer	273	2,287	2,239	1,020	0	1,309	3,129	8,891	9,878

Table A.1—Continued

Code	Description	2000	2001	2002	2003	2004	2005	2006	2007	2008
15K	Aircraft Components Repair Supervisor	N/A	N/A	N/A	N/A	N/A	N/A	3,924	11,413	13,840
15M	Uh–1 Helicopter Repairer	822	N/A	N/A	N/A	N/A	N/A	N/A	N/A	N/A
15N	Avionic Mechanic	556	783	40	54	98	1,171	1,136	6,320	5,326
15P	Aviation Operations Specialist	431	439	21	33	444	1,046	844	7,241	4,724
15Q	Air Traffic Control Operator	991	1,620	1,158	536	134	2,225	2,082	6,090	8,598
15R	Ah–64 Attack Helicopter Repairer	771	223	0	11	101	1,737	1,650	7,836	3,718
15S	Oh–58D Helicopter Repairer	1,777	1,312	306	0	0	2,106	2,287	7,029	4,615
15T	Uh–60 Helicopter Repairer	2,147	1,614	1,620	661	718	2,252	2,054	5,692	2,584
15U	Ch–47 Helicopter Repairer	1,683	3,199	2,015	487	1,079	2,354	2,205	5,193	2,168
15W	Unmanned Aerial Vehicle Operator	1,504	768	61	334	275	1,498	1,697	7,226	3,754
15X	Ah–64A Armament/Electrical Systems Repairer	1,706	421	0	311	N/A	N/A	2,212	7,573	6,211
15Y	Ah–64D Armament/Electrical/ Avionic System Repairer	304	1,374	2,477	2,565	2,949	3,021	2,219	8,084	4,084
18X	Special Forces Recruit	N/A	N/A	11,293	8,637	7,777	14,705	16,465	21,611	34,127
19D	Cavalry Scout	3,012	3,352	3,297	1,472	1,803	7,713	7,988	7,006	5,335
19K	M1 Armor Crewman	2,797	4,839	4,301	2,775	2,477	4,600	3,334	6,997	9,155
21B	Combat Engineer	2,307	1,217	1,493	1,410	1,199	4,265	3,863	4,817	5,927
21C	Bridge Crewmember	2,474	455	378	673	1,580	4,233	3,973	4,306	9,822
21D	Diver	2,534	867	0	59	71	2,834	2,202	12,158	11,288
21E	Heavy Construction Equipment Operator	598	793	504	286	725	1,413	1,584	4,654	4,623
21K	Plumber	451	N/A	222	0	N/A	773	1,931	3,594	3,142
21M	Firefighter	417	548	0	0	0	1,764	1,809	4,572	2,618
21R	Interior Electrician	259	591	132	N/A	N/A	1,955	1,157	5,572	3,167
21T	Technical Engineer	1,783	1,023	28	0	0	2,543	2,844	9,912	4,271
21W	Carpentry and Masonry Specialist	413	336	667	99	202	1,954	2,306	3,500	1,811
21Y	Topographic Engineering Supervisor	734	759	26	1,048	1,387	3,721	6,854	13,126	18,927
25B	Information Systems Operator–Analyst	1,230	928	46	45	561	2,533	3,539	9,990	6,231
25C	Radio Operator–Maintainer	885	216	402	305	194	N/A	9,020	8,758	6,780
25F	Network Switching Systems Operator–Maintainer	2,788	3,507	2,330	938	2,243	8,565	7,418	10,836	7,245
25L	Cable Systems Installer–Maintainer	446	429	344	1,101	537	2,594	2,409	8,695	5,354
25M	Multimedia Illustrator	1,670	1,757	137	105	655	2,470	3,632	7,425	2,117
25N	Nodal Network Systems	N/A	N/A	N/A	N/A	N/A	N/A	N/A	5,447	13,207
25P	Microwave Systems Operator–Maintainer	5,144	3,416	4,409	5,130	5,448	7,924	11,791	22,045	33,929
25Q	Multichannel Transmission Systems Operator–Maintainer	4,442	4,861	2,727	1,569	1,612	4,166	6,248	10,411	26,725

Table A.1—Continued

Code	Description	2000	2001	2002	2003	2004	2005	2006	2007	2008
25R	Visual Information Equipment Operator–Maintainer	N/A	N/A	119	0	N/A	N/A	5,585	9,510	9,458
25S	Satellite Communication Systems Operator–Maintainer	3,338	3,754	4,011	5,426	6,541	14,159	13,531	19,309	35,931
25U	Signal Support Systems Specialist	1,485	654	55	733	1,008	4,834	3,706	12,782	24,022
25V	Combat Documentation/Production Specialist	1,523	1,315	34	0	N/A	1,972	2,253	6,562	2,055
27D	Paralegal Specialist	1,481	1,931	118	18	0	3,298	3,665	11,824	13,514
31B	Military Police	755	883	1,615	275	132	3,843	3,341	6,297	9,377
31E	Internment/Resettlement Specialist	1,477	1,401	1,198	131	264	6,418	11,648	9,746	6,155
35F	Intelligence Analyst	2,856	1,648	119	275	1,038	5,658	5,184	10,730	14,171
35G	Imagery Analyst	2,470	2,443	601	1,255	2,004	3,431	3,899	11,637	16,666
35H	Common Ground Station (CGS) Operator	6,815	6,676	4,738	565	0	N/A	1,150	N/A	N/A
35M	Human Intelligence Collector	3,224	4,816	810	1,958	2,128	6,374	4,681	3,159	N/A
35S	Signals Collector/Analyst	N/A	N/A	N/A	N/A	N/A	N/A	2,444	2,946	N/A
35T	Military Intelligence Systems Maintainer/Integrator	5,697	6,485	3,918	6,017	3,801	4,144	6,906	9,957	N/A
35V	Translator Aide	1,496	1,072	845	293	0	N/A	N/A	N/A	N/A
35W	EW/Signt Recruit	9,619	12,295	13,553	14,289	15,040	14,715	12,716	11,293	N/A
35Z	Signals Intelligence (Electronic Warfare)/Senior Sergeant/Chief	5,247	4,582	3,469	2,530	1,168	2,935	3,259	6,294	N/A
37F	Psychological Operations Specialist	2,617	3,398	422	197	386	5,028	10,279	17,518	17,215
42A	Human Resources Specialist	697	589	75	14	115	2,627	1,201	5,798	2,005
42F	Human Resources Information Systems Management Specialist	1,992	1,802	213	10	41	4,081	2,842	13,941	1,829
44B	Metal Worker	217	58	0	46	0	1,997	538	4,925	4,646
44C	Financial Management Technician	769	1,615	36	8	19	4,356	3,507	7,476	5,378
45B	Small Arms/Artillery Repairer	235	450	351	280	46	1,169	1,383	4,780	2,685
45G	Fire Control Repairer	398	280	289	454	1,660	3,197	5,009	9,575	9,823
45K	Armament Repairer	1,467	N/A	1,307	1,532	874	6,225	3,111	2,807	5,925
46Z	Chief Public Affairs NCO	1,927	2,262	127	0	246	5,304	5,288	11,509	9,694
52C	Utilities Equipment Repairer	1,054	729	799	335	579	1,823	3,891	2,856	3,811
52D	Power–Generation Equipment Repairer	4,413	2,408	3,363	2,896	3,714	6,887	4,221	6,297	9,223
56M	Chaplain Assistant	1,175	1,393	1,005	237	104	1,930	3,188	8,105	3,040
63A	Abrams Tank System Maintainer	830	2,741	1,691	570	1,278	3,297	2,253	4,077	7,647
63B	Light–Wheel Vehicle Mechanic	2,114	2,344	1,014	681	825	N/A	N/A	N/A	N/A
63D	Artillery Mechanic	281	971	1,507	817	N/A	N/A	N/A	N/A	N/A
63H	Track Vehicle Repairer	4,185	4,139	3,568	1,781	3,316	N/A	N/A	N/A	N/A

Table A.1—Continued

Code	Description	2000	2001	2002	2003	2004	2005	2006	2007	2008
63J	Quartermaster and Chemical Equipment Repairer	304	312	1,435	2,239	1,752	4,958	9,275	9,078	18,050
63M	Bradley Fighting Vehicle System Maintainer	948	3,914	2,698	3,431	2,427	6,630	4,961	6,449	10,407
63X	Vehicle Maintenance Supervisor	361	589	2,028	1,573	1,518	4,005	3,590	7,375	4,689
63Z	Mechanical Maintenance Supervisor	2,550	1,377	804	1,052	2,232	6,640	7,142	6,216	6,604
68A	Biomedical Equipment	2,553	1,063	N/A	0	55	3,901	3,901	8,888	2,494
68D	Operating Room Specialist	536	860	0	0	43	3,320	2,349	N/A	N/A
68E	Dental Specialist	508	378	0	21	102	1,771	1,382	6,576	4,574
68G	Patient Administration Specialist	381	0	0	0	0	1,622	N/A	N/A	N/A
68H	Optical Laboratory Specialist	N/A	526	0	0	N/A	N/A	N/A	N/A	N/A
68J	Medical Logistics Specialist	508	72	0	0	0	2,423	1,636	6,372	3,181
68K	Medical Laboratory Specialist	2,644	5,994	6,140	4,642	5,823	5,540	N/A	N/A	N/A
68M	Nutrition Care Specialist	694	616	0	0	0	2,974	2,025	10,686	2,516
68P	Radiology Specialist	1,409	936	12	0	179	2,265	N/A	N/A	N/A
68Q	Pharmacy Specialist	356	1,032	884	1,617	521	2,097	2,894	6,721	5,722
68R	Veterinary Food Inspection Specialist	803	810	638	132	49	1,578	1,520	5,745	6,286
68S	Preventive Medicine Specialist	1,618	2,635	1,584	138	35	1,710	8,141	N/A	N/A
68T	Animal Care Specialist	784	832	0	129	42	2,461	1,925	7,863	5,635
68W	Health Care Specialist	975	889	981	1,033	105	2,210	7,654	12,571	13,612
68X	Mental Health Specialist	1,731	2,092	0	0	167	2,129	3,498	N/A	N/A
74D	Chemical, Biological, Radiological and Nuclear (CBRN) Specialist	1,834	1,390	1,425	1,530	1,950	5,933	4,483	8,044	10,293
88H	Cargo Specialist	340	575	410	303	725	4,132	2,898	8,290	5,994
88K	Watercraft Operator	713	1,096	610	330	0	2,543	3,022	2,475	5,059
88L	Watercraft Engineer	834	471	485	919	1,675	N/A	515	2,968	375
88M	Motor Transport Operator	715	1,540	927	2,083	3,429	8,730	10,314	11,039	18,923
88N	Transportation Management Coordinator	1,104	241	0	0	18	2,177	736	4,360	2,153
89A	Ammunition Stock Control and Accounting Specialist	N/A	N/A	N/A	N/A	N/A	N/A	N/A	N/A	11,728
89B	Ammunition Specialist	2,605	2,529	274	358	1,141	5,954	8,200	11,515	10,655
89D	Explosive Ordnance Disposal Specialist	2,469	6,409	3,355	2,651	5,448	13,236	15,659	19,239	28,400
92A	Automated Logistical Specialist	913	2,261	387	644	945	2,602	2,041	6,234	3,669
92F	Petroleum Supply Specialist	1,927	4,603	4,539	4,907	5,397	9,600	15,282	11,984	22,481
92G	Food Service Specialist	2,594	4,712	4,228	4,467	4,659	9,163	8,604	10,285	16,592
92L	Petroleum Laboratory Specialist	1,007	N/A	450	165	738	2,760	1,580	10,489	4,571
92M	Mortuary Affairs Specialist	202	199	0	0	0	1,758	1,741	5,302	4,816
92R	Parachute Rigger	2,723	4,191	3,845	2,079	2,304	7,829	6,885	7,781	16,903

Table A.1—Continued

Code	Description	2000	2001	2002	2003	2004	2005	2006	2007	2008
92S	Shower/Laundry and Clothing Repair Specialist	93	37	72	68	173	1,637	1,410	4,772	4,175
92W	Water Treatment Specialist	208	740	510	340	1,697	5,802	6,111	7,481	7,348
92Y	Unit Supply Specialist	1,050	1,491	550	396	674	2,308	2,414	5,795	5,248
94A	Land Combat Electronic Missile System Repairer	2,270	732	108	0	1,444	5,943	7,295	11,862	27,318
94E	Radio and Communications Security (Comsec) Repairer	5,113	3,736	2,335	2,783	4,263	8,454	18,409	16,654	29,637
94F	Special Electronic Devices Repairer	1,499	538	389	347	2,122	9,376	12,016	16,708	13,769
94H	Test, Measurement, and Diagnostic Equipment (TMDE) Maintenance Support Specialist	1,805	N/A	2,262	1,659	N/A	N/A	7,507	13,163	19,604
94L	Avionic Communications Equipment Repairer	1,040	N/A	44	N/A	169	2,734	2,389	12,252	5,143
94M	Radar Repairer	1,942	1,534	620	1,267	1,673	6,617	11,827	13,663	7,188
94P	Multiple Launch Rocket System Repairer	405	174	414	1,008	629	1,748	4,933	6,412	2,314
94R	Avionic System Repairer	353	178	556	975	0	N/A	1,956	9,034	5,647
94T	Avenger System Repairer	619	733	517	1,115	1,382	N/A	N/A	18,524	16,781
94Y	Integrated Family of Test Equipment (IFTE) Operator and Maintainer	N/A	N/A	N/A	N/A	N/A	N/A	4,336	11,075	26,869
94Z	Senior Electronic Maintenance Chief	1,265	1,073	939	96	1,396	2,780	6,365	15,804	22,401
96R	Ground Surveillance Systems Operator	1,319	945	296	0	0	N/A	N/A	N/A	N/A

NOTES: Where possible we have recoded obsolete occupational specialties to reflect the classification system that was current as of October 2008. Amounts are in FY 2008 dollars.

Table A.2
Average Bonuses for Army Enlistees, by Length of Enlistment Term, FY 2000–2008

Term	2000	2001	2002	2003	2004	2005	2006	2007	2008
2 years	1,013	1,602	2,036	1,842	1,795	2,325	2,227	12,485	12,188
3 years	593	1,533	1,099	720	1,324	5,462	5,483	5,390	9,330
4 years	2,810	4,146	3,129	1,798	2,618	7,359	8,849	11,375	14,889
6 years	3,722	4,978	5,306	4,428	4,554	7,867	9,454	12,038	16,414

NOTE: Amounts are in FY 2008 dollars.

Table A.3
Average Bonuses for Navy Enlistees, by Enlistment Classification, FY 1999–2008

Code	Description	1999	2000	2001	2002	2003	2004	2005	2006	2007	2008
AB	Aviation Boatswain's Mate	355	585	524	838	198	999	1,110	4,406	2,178	1,833
AC	Air Traffic Controller	622	853	418	731	359	344	367	0	0	342
AD	Aviation Machinist's Mate	273	472	490	700	94	199	362	3,359	1,832	1,628
ADEK	Aviation Deck	N/A	N/A	N/A	N/A	N/A	N/A	N/A	N/A	254	1,053
ADMN	Administration	N/A	2,949	810	1,978	2,869	4,142	N/A	N/A	N/A	N/A
ADSP	Adminstration and Support	N/A	N/A	N/A	N/A	N/A	N/A	N/A	N/A	0	88
AE	Aviation Electrician's Mate	3,964	5,560	6,780	7,568	N/A	N/A	N/A	N/A	N/A	N/A
AECF	Advanced Electronics/Computer	N/A	N/A	4,284	4,560	4,308	5,275	4,295	5,227	5,359	6,243
AG	Aerographer's Mate	343	595	508	182	386	621	997	59	1,704	1,212
AIC	Air Intercept Controller	35	831	2,278	4,244	N/A	N/A	N/A	N/A	N/A	N/A
AIR	Aircrew	306	722	2,166	3,124	4,662	4,370	1,704	4,389	6,931	8,525
AK	Aviation Storekeeper	388	825	748	1,497	N/A	N/A	N/A	N/A	N/A	N/A
AM	Aviation Structural Mechanic	499	1,313	2,107	3,206	3,540	3,027	1,950	3,266	1,416	1,608
AMEK	Aviation Mechanical	N/A	N/A	N/A	N/A	N/A	N/A	N/A	N/A	802	1,521
AN	Airman	1,179	2,095	1,011	892	1,431	1,920	1,565	N/A	N/A	0
AO	Aviation Ordnanceman	556	694	970	1,655	2,313	4,002	1,147	4,339	1,970	1,617
AS	Aviation Support Equipment Technician	316	978	704	477	87	201	599	3,400	1,436	1,623
AT	Aviation Electronics Technician	252	781	1,462	N/A	N/A	N/A	N/A	N/A	N/A	N/A
AV	Aviation Avionics Technician	N/A	0	1,516	3,939	3,523	3,560	2,305	3,668	1,761	1,707
AW	Aviation Warfare Systems Operator	N/A	N/A	N/A	N/A	N/A	N/A	N/A	N/A	N/A	N/A
AZ	Aviation Maintenance Administrator	426	0	436	192	370	362	337	0	0	333
BM	Boatswain's Mate	N/A	2,907	N/A	1,230	2,463	3,374	N/A	0	0	0
BU	Builder	265	0	0	86	145	184	362	1,790	1,514	1,858
CE	Construction Electrician	109	14	80	43	305	193	211	39	1,558	2,242
CM	Construction Mechanic	198	15	30	628	2,220	4,278	624	770	1,648	2,023
CONT	Controller/Guidance	N/A	N/A	N/A	N/A	3,399	N/A	N/A	N/A	N/A	N/A
CS	Culinary Specialist	N/A	N/A	N/A	N/A	N/A	N/A	N/A	33	3	0
CT	Cryptologic Technician	1,064	1,497	2,859	4,147	2,894	2,283	4,782	3,158	5,072	5,731
DC	Damage Controlman	280	823	427	170	371	1,898	1,968	3,788	1,670	1,643
DK	Disbursing Clerk	318	822	871	1,612	676	943	N/A	N/A	N/A	N/A
DT	Dentalman	10	0	0	262	236	365	711	N/A	N/A	N/A
EA	Engineering Aide	78	0	862	1,521	N/A	582	1733	69	1,366	1,934

Table A.3—Continued

Code	Description	1999	2000	2001	2002	2003	2004	2005	2006	2007	2008
EL4	Electronics, 4-year	N/A	N/A	N/A	N/A	N/A	N/A	N/A	N/A	954	1,069
EL6	Electronics, 6-year	N/A	N/A	N/A	N/A	N/A	N/A	N/A	N/A	5,407	6,226
ELCL	Electrical	N/A	3,163	1,136	1,090	2,500	N/A	N/A	N/A	N/A	N/A
ELCT	Electronics	N/A	N/A	N/A	N/A	3,336	N/A	N/A	N/A	N/A	N/A
EM	Electrician's Mate	261	375	541	1,188	314	1,900	3,160	3,759	1,808	1,584
EN	Engineman	415	941	1,608	1,706	311	1,480	2,747	3,632	1,605	1,631
ENGR	Engineering	N/A	2,887	611	974	2,180	3,307	N/A	N/A	N/A	N/A
EO	Equipment Operator	42	0	400	834	176	70	308	1,683	1,350	2,452
EOD	Explosive Ordnance Disposal	N/A	N/A	N/A	N/A	N/A	N/A	N/A	29,839	37,191	38,643
ET	Electronics Technician	4,575	N/A	N/A	N/A	N/A	N/A	N/A	N/A	N/A	N/A
EW	Electronic Warfare Technician	369	2,059	5,497	4,943	6,128	N/A	N/A	N/A	N/A	N/A
FN	Fireman	1,926	2,854	1,026	949	657	2,046	N/A	N/A	N/A	N/A
FT	Fire Control Technician	6,926	N/A	N/A	N/A	N/A	N/A	N/A	N/A	N/A	N/A
GM	Gunner's Mate	487	489	878	1,430	955	4,587	3,165	150	0	0
GS	Gas Turbine System Technician	36	33	697	2,509	2,862	3,420	898	3,821	1,995	1,699
HCMB	Hull/Combat Systems	N/A	N/A	N/A	1,211	2,528	N/A	N/A	N/A	N/A	N/A
HM	Hospital Corpsman	5	12	18	369	1,075	1,502	616	61	3	159
HT	Hull Maintenance Technician	395	1,177	1,976	4,175	1,276	2,788	2,138	3,257	1,127	1,758
IC	Interior Communications Electrician	265	484	1,705	1,858	755	1,429	1,227	3,455	1,001	1,437
IS	Intelligence Specialist	416	745	1,469	1,790	1,222	1,295	4,347	3,900	3,741	4,104
IT	Information Systems Technician	1,341	1,042	557	1,218	845	3,107	3,763	3,723	2,787	1,889
JO	Journalist	321	290	27	399	2,186	1,596	1,253	N/A	N/A	N/A
LI	Lithographer	169	0	0	635	272	621	231	N/A	N/A	N/A
MA	Machine Accountant	N/A	591	1,225	2,583	2,051	2,801	765	195	2,564	2,384
MC	Mass Communications Specialist	N/A	N/A	N/A	N/A	N/A	N/A	N/A	0	0	0
MCHA	Mechanical/Aviation	N/A	3,079	N/A	1,151	2,570	3,490	N/A	N/A	N/A	N/A
MED	Medical	N/A	2,804	311	1,368	2,605	3,716	N/A	N/A	N/A	N/A
MM	Machinist's Mate	379	1,489	2,626	2,591	1,870	2,267	2,656	3,856	3,066	3,272
MN	Mineman	1,745	2,023	4,425	2,826	1,404	3,287	1,766	2,900	2,043	1,761
MR	Machinery Repairman	238	608	2,175	6,208	618	1,671	923	3,434	1,542	932
MS	Mess Management Specialist	456	1,334	1,649	2,994	3,715	4,980	225	185	24	N/A
MT	Missile Technician	5,349	6,836	6,903	6,682	1,142	4,995	5,376	5,719	5,724	5,900
MU	Musician	193	1,292	3,433	5,050	4,234	4,077	5,144	228	3,351	2,973

Table A.3—Continued

Code	Description	1999	2000	2001	2002	2003	2004	2005	2006	2007	2008
NAV	Navigation	N/A	2,914	1,111	1,057	2,339	N/A	N/A	N/A	N/A	N/A
ND	Navy Diver	N/A	N/A	N/A	N/A	N/A	N/A	N/A	25,262	33,014	32,340
OPCM	Operations/Communications	N/A	2,842	N/A	1,248	2,700	3,367	N/A	N/A	N/A	N/A
ORDN	Ordnance	N/A	N/A	N/A	834	2,846	N/A	N/A	N/A	N/A	N/A
OS	Operations Specialist	325	841	412	914	1,714	2,478	1,083	38	0	0
PC	Postal Clerk	493	1,222	N/A	271	2,128	3,093	3,812	3,368	1,836	2,162
PH	Photographer's Mate	207	0	57	700	1,815	1,901	1,813	468	N/A	N/A
PN	Personnelman	323	784	923	1,332	665	568	241	N/A	N/A	N/A
PR	Aircrew Survival Equipmentman	616	1,077	1,427	2,962	2,257	4,201	1,181	96	0	0
PS	Personnel Specialist	N/A	N/A	N/A	N/A	N/A	N/A	857	0	0	116
QM	Quartermaster	544	906	573	690	561	3,038	1,382	155	0	0
RM	Radioman	1,831	N/A	N/A	N/A	N/A	N/A	N/A	N/A	N/A	N/A
RP	Religious Programs Specialist	208	686	0	271	827	579	706	0	0	185
SB	Special Warfare Boat Operator	N/A	N/A	N/A	N/A	N/A	N/A	N/A	17,923	28,911	33,249
SEC	Submarine Electronics/Computer	3,629	5,864	6,656	6,342	2,168	5,361	5,647	5,563	6,241	6,600
SENG	Surface Engineering	N/A	N/A	N/A	N/A	N/A	N/A	N/A	1,444	2,067	2,803
SH	Ship's Serviceman	489	1,148	1,221	3,031	338	1,600	218	0	0	197
SK	Storekeeper	363	1,205	1,249	1,950	1,736	2,949	1,130	3,508	1,228	204
SN	Seaman	1,033	2,014	998	1,032	1,146	1,886	1,543	6,256	0	0
SO	Sonarman	N/A	N/A	N/A	N/A	N/A	N/A	N/A	35,900	36,837	38,124
SPSV	Special Services	N/A	3,352	N/A	1,071	2,288	3,363	N/A	N/A	N/A	N/A
SS	Seaman Submarine Program	1,361	2,589	2,460	2,698	1,697	N/A	N/A	N/A	N/A	N/A
ST	Sonar Technician	947	1,518	1,624	2,230	770	1,266	3,808	4,503	3,287	3,531
SW	Steelworker	196	0	190	28	39	1,496	535	0	0	129
TM	Torpedoman's Mate	355	159	137	371	288	669	320	0	N/A	N/A
UT	Utilitiesman	407	0	426	755	2,508	171	273	1,751	1,485	2,686
YN	Yeoman	346	475	794	588	545	485	464	137	0	87

NOTES: Amounts are in FY 2008 dollars. Classifications with fewer than 30 enlistments in a particular fiscal year are not listed in the table. The majority of new enlistees enlist into specific occupations called ratings. However, through several programs such as the GTEP (GENDET Targeted Enlistment Program), the Navy also permits individuals to enlist into a generic grouping of occupations at the time of enlistment and then choose a specific occupation later during their term of service. The table includes both individuals who enlist in specific occupations as well as those who enlist into broader occupational groupings and follows the enlistee classification system developed by the Navy. With the exception of Explosive Ordnance Disposal (EOD) and Air Intercept Controller (AIC), both of which refer to specific occupations, all classifications with codes of three or more letters are generic, and all of the two letter codes reflect specific ratings.

Table A.4
Average Bonuses for Navy Enlistees, by Length of Enlistment Term, FY 1999–2008

Term of Service	1999	2000	2001	2002	2003	2004	2005	2006	2007	2008
4 years	335	1,115	415	598	1,107	1,341	641	1	0	0
5 years	1,482	2,026	2,783	3,143	2,756	3,783	2,852	3,429	2,899	3,024
6 years	4,538	6,016	6,823	6,184	5,914	5,656	5,944	8,728	14,540	14,511

NOTE: Amounts are in FY 2008 dollars.

Detailed Background on Reenlistment Bonuses

Table B.1
SRB Multipliers for Selected Skills, by MOS and Grade, June 2007

		Multiplier							
		Grade							
		E-4	E-5	E-5	E-5	E-6	E-6	E-6	
		Zone							
MOS	SQI (in Selected MOSs)	A	A	B	C	A	B	C	Bonus Cap (Zones)
All MOS	Special Forces (T)	0	4	4	4	4	4	4	$50K(A,B,C)
11B		1	1.5	2	0	1.5	2	2	$15K(A),$25K(B,C)
11B	Ranger (G), Airborne Ranger (V)	1.5	2	2.5	0	2	2.5	0	$15K(A),$25K(B,C)
11C		1.5	1.5	2	0	1.5	2	2	$15K(A),$25K(B,C)
13B	Parachute Qualified (P)	0	1	1	0	0	0	0	$10K(A),$15K(B,C)
13D		1.5	1	1	0	1	1	0	$10K(A),$20K(B,C)
13D	Parachute Qualified (P)	2.5	2.5	2.5	0	1.5	2	0	$10K(A),$20K(B,C)
13F		1.5	2	2	0	2	2	0	$10K(A),$20K(B,C)
18B		2.5	2.5	2.5	2.5	2.5	2.5	2.5	$30K(A,B,C)
18C		2.5	2.5	2.5	2.5	2.5	2.5	2.5	$30K(A,B,C)
18D		3.5	3.5	3.5	3.5	3.5	3.5	3.5	$30K(A,B,C)
18E		3.5	3.5	3.5	3.5	3.5	3.5	3.5	$30K(A,B,C)
18F		0	0	0	0	0	3.5	3.5	$30K(A,B,C)
19D		1	1	1	0	1	1	0	$10K(A) $20K(B,C)
19D		0	2	2	0	0	0	0	$10K(A) $20K(B,C)
21C		0	1.5	1.5	0	0	0	0	$10K(A) $20K(B,C)
21D		2	2	2	0	0	2	2	$15K(A) $25K(B,C)
21E		1	1	0	0	0	0	0	$10K(A) $15K(B,C)
21J		1	1	0	0	0	0	0	$10K(A) $15K(B,C)
21J		0	1.5	1.5	0	0	0	0	$10K(A) $15K(B,C)
21K		1.5	1	0	0	0	0	0	$10K(A) $15K(B,C)
21P		1.5	1.5	1.5	0	1.5	1.5	1.5	$15K(A,B,C)
21R		2	1	0	0	0	0	0	$15K(A,B,C)

Table B.1—Continued

		Multiplier							
		Grade							
		E-4	E-5	E-5	E-5	E-6	E-6	E-6	
		Zone							
MOS	SQI (in Selected MOSs)	A	A	B	C	A	B	C	Bonus Cap (Zones)
21U		1	1	1	0	0	0	0	$10K(A) $20K(B,C)
21W		1	1	0	0	0	0	0	$10K(A) $15K(B,C)
25B		1	1	1	0	1	1	0	$10K(A) $20K(B,C)
25B		1.5	1.5	1.5	0	0	0	0	$10K(A) $20K(B,C)
25C		1	0.5	0	0	0	0	0	$10K(A) $15K(B,C)
25D		1.5	1.5	1.5	0	1.5	1.5	0	$10K(A) $15K(B,C)
25D		2	0	0	0	0	0	0	$10K(A) $15K(B,C)
25P		1.5	1	1	0	1	1	0	$10K(A) $15K(B,C)
25Q		1	1	1	0	0	0	0	$10K(A) $15K(B,C)
25Q		0	1.5	1.5	0	0	0	0	$10K(A) $15K(B,C)
25S		3	2.5	2	0	2.5	2	0	$30K(A,B,C)
25U		1.5	1.5	1.5	0	0	0	0	$10K(A) $15K(B,C)
25U		2.5	2.5	2.5	0	0	0	0	$10K(A) $15K(B,C)
27D		0.5	0	0	0	0	0	0	$10K(A) $15K(B,C)
27D		0	0.5	0	0	0.5	0	0	$10K(A) $15K(B,C)
31B		1	1.5	1.5	0	1	1	1	$10K(A) $15K(B,C)
31D		0	1	1	0	0	0	0	$10K(A) $15K(B,C)
31E		1.5	1.5	1.5	0	0.5	0.5	0.5	$10K(A) $20K(B,C)
33W		1	2	2	0	0	0	0	$15K(A) $20K(B,C)
37F		2	2	2	2	1.5	1.5	1	$20K(A,B,C)
38B		0	0.5	0.5	0	1.5	1.5	0	$10K(A) $20K(B,C)
44B		2.5	1.5	1	0	0	0	0	$10K(A) $15K(B,C)
45B		1.5	1.5	1	0	0	0	0	$10K(A) $15K(B,C)
45G		1	0.5	0	0	0	0	0	$10K(A) $15K(B,C)
46Q		1	1	1	0	1	1.5	0	$10K(A) $15K(B,C)
46R		2.5	2.5	0	0	0	0	0	$15K(A,B,C)
56M		0.5	0	0	0	0	0	0	$10K(A) $15K(B,C)
56M		0	0.5	0	0	0	0	0	$10K(A) $15K(B,C)
63B		1	1	1	0	1	1	0	$10K(A) $20K(B,C)

Table B.1—Continued

MOS	SQI (in Selected MOSs)	E-4	E-5	E-5	E-5	E-6	E-6	E-6	Bonus Cap (Zones)
		Multiplier							
		Grade							
		Zone							
		A	A	B	C	A	B	C	
63B		2	2	2	0	2	2	0	$10K(A) $20K(B,C)
63J		0.5	0.5	0.5	0	0	0	0	$10K(A) $20K(B,C)
63J		2	2	1.5	0	0	0	0	$10K(A) $20K(B,C)
68E		1	0	0	0	0	0	0	$10K(A) $15K(B,C)
68J		0	0	0	0	1	1	0	$10K(A) $15K(B,C)
68K		1	0	0	0	0	0	0	$10K(A) $15K(B,C)
68K		1.5	1.5	1.5	0	1.5	1.5	0	$20K(A,B,C)
68S		1	0	0	0	0	0	0	$10K(A) $15K(B,C)
68T		0.5	0.5	0	0	0	0	0	$10K(A) $15K(B,C)
68W		1	1	1.5	0	1	1.5	0	$15K(A) $20K(B,C)
68W		2	2	2	0	0	0	0	$20K(A,B,C)
68W		2	2	2	0	0	0	0	$20K(A,B,C)
74D		0	2.5	2.5	0	0	0	0	$10K(A) $5K(B)
79R		0	1	1	1.5	1	1	1.5	$10K(A) $20K(B,C)
88M		2	1	1.5	0	1	1	0	$10K(A) $20K(B,C)
88M		0	2	2	0	2	2	0	$10K(A) $20K(B,C)
88N		1.5	0	0	0	0	0	0	$10K(A) $15K(B,C)
89B		1	1	1	0	1	1	0	$10K(A) $20K(B,C)
89B		2.5	2.5	2	0	0	0	0	$10K(A) $20K(B,C)
89D		2.5	3.5	3.5	0	3	4	4	$40K(A,B,C)
92A		1.5	1.5	2	0	0	0	0	$10K(A) $15K(B,C)
92F		0.5	1	1	0	0	0	0	$10K(A) $20K(B,C)
92F		0	2	1.5	0	1	1	0	$10K(A) $20K(B,C)
92G		1	1	1.5	0	0	0	0	$10K(A) $15K(B,C)
92L		3.5	0	0	0	0	0	0	$10K(A) $15K(B,C)
92R		1.5	1	0	0	0	0	0	$10K(A) $15K(B,C)
92W		2.5	2.5	1.5	0	0	0	0	$10K(A) $15K(B,C)
92Y		0.5	0	0	0	0	0	0	$10K(A) $15K(B,C)
92Y		1	1	1	0	0	0	0	$10K(A) $15K(B,C)

Table B.1—Continued

		Multiplier							
		Grade							
		E-4	E-5	E-5	E-5	E-6	E-6	E-6	
		Zone							
MOS	SQI (in Selected MOSs)	A	A	B	C	A	B	C	Bonus Cap (Zones)
94A		2	0.5	0.5	0	0	0	0	$10K(A) $15K(B,C)
94A		1.5	1.5	1.5	0	0	0	0	$10K(A) $15K(B,C)
94E		1.5	0	0	0	0	0	0	$15K(A,B,C)
94F		1.5	1	1	0	1	1	0	$15K(A) $20K(B,C)
94F		2.5	2	2	0	2	0	0	$15K(A) $20K(B,C)
94H		1.5	1.5	1.5	3	1.5	1.5	3	$10K(A) $20K(B,C)
94M		2	0	0	0	0	0	0	$15K(A,B,C)
94S		2	2	2	2	2	2	2	$15K(A) $20K(B,C)
94T		2	0.5	0	0	0	0	0	$15K(A,B,C)
94Y		2	0	0	0	0	0	0	$15K(A,B,C)
96B		1	2	2	0	2	2	2	$10K(A) $15K(B,C)
96B		2.5	3.5	4	0	3.5	4	0	$10K(A) $15K(B,C)
96D		2.5	1.5	1.5	0	1.5	1.5	0	$20K(A,B,C)
96D		3.5	2.5	3	0	2	3	0	$20K(A,B,C)
96H		2.5	1.5	1.5	0	1.5	1.5	0	$20K(A,B,C)
96U		1	1	0	0	0	0	0	$15K(A,B,C)
96U		0	2	0	0	0	0	0	$15K(A,B,C)
97B		0	0	0	0	2	2	2	$20K(A,B,C)
97E		4	4	4	4.5	4	4	4.5	$30K(A,B,C)
97E		4.5	4.5	4.5	0	4.5	4.5	0	$30K(A,B,C)
98C		2	2	2	0	2	2	0	$15K(A,B,C)
98C		2.5	2.5	2.5	0	2	2.5	0	$15K(A,B,C)
98G		0	2	3	0	2	3	1.5	$30K(A,B,C)
98G		2	2.5	3.5	0	2.5	3.5	0	$30K(A,B,C)
98G		1.5	2	2	0	2	2	0	$10K(A) $20K(B,C)
98G		1	1.5	1.5	0	1.5	1.5	0	$10K(A) $20K(B,C)
98G		1	1.5	1.5	0	1.5	1.5	0	$10K(A) $20K(B,C)
98Y		1.5	1.5	1.5	0	0	0	0	$15K(A,B,C)

SOURCE: Milpers Message Number 07-141, Army Human Resources Command, June 6, 2007.

NOTE: A, B, and C in this table refer to reenlistment Zones A, B, and C.

Table B.2
Skills Eligible for the Enhanced SRB, December 2007

MOS	MOS Title	Number in Zones A, B, and C	Critical Skill	Special Critical Skill	Location Critical Skill	SQI, Unit
11B	Infantryman	37,414	Yes		Yes	G/V, 75th Ranger
11C	Ind Fire Infantryman	4,246	Yes			
13B	Cannon Crewmember	6,168	Yes			
13D	Field Artillery Data System Specialist	1,977	Yes			
13F	Fire Support Specialist	4,456	Yes		Yes	G/V, 75th Ranger
13R	Field Artillery Radar Operator	423	Yes			
13W	Field Artillery Crewmember	251	Yes			
14E	Patriot Fire Operator/Maintainer	910	Yes			
14J	Tactical Operations Operator/Maintainer	1,228	Yes			
14S	Avenger Crew	648	Yes			
14T	Patriot Operator/Maintainer	1,442	Yes			
15F		353			Yes	P, 160TH SOAR
15G		480			Yes	P, 160TH SOAR
15Q	Air Traffic Control	691	Yes			
15T		2,760			Yes	P, Special Operations Command
18B	Specialist for Weapons Sergeant	907		Yes		
18C	Specialist for Engineering Sergeant	868		Yes		
18D	Specialist for Med Sergeant	759		Yes		
18E	Specialist for Communications Sergeant	922		Yes		
18F	Specialist for Intelligence Sergeant	118		Yes		
19D	Cavalry Scout	8,175	Yes			
19K	M1 Armor Crewman	5,316	Yes			
21C	Bridge Crewmember	429	Yes			
21D	Diver	103	Yes			
21E	Construction Equipment Operator	983	Yes			
21J	General Construction Equipment Operator	610	Yes			
21K	Plumber	200	Yes			
21P	Prime Power Specialist	153	Yes			
21U	Topographic Analyst	438	Yes			
25B	Information Technology Specialist	4,453	Yes		Yes	P, Special Operations Command

Table B.2—Continued

MOS	MOS Title	Number in Zones A, B, and C	Critical Skill	Special Critical Skill	Location Critical Skill	SQI, Unit
25C	Radio Operator/Maintainer	731	Yes		Yes	G/V, 75TH Ranger
25P	Microwave Operator/Maintainer	529	Yes			
25Q	Multich System Operator/Maintainer	1,830	Yes			
25S	Satellite Communication System Operator/Maintainer	1,905		Yes		
25U	Signal Support System Specialist	5,251	Yes		Yes	G/V, 75TH Ranger
27D	Paralegal Specialist	818	Yes		Yes	G/V, 75TH Ranger
31B	Military Police	9,469	Yes			
31D	CID Special Agent	414	Yes			
31E	Intern/Resettle Specialist	914	Yes			
35F	Intelligence Analyst	3,806		Yes	Yes	G/V, 75TH Ranger
35G	Imagery Analyst	554		Yes		
35H	Common Ground Station Operator	350		Yes		
35K	Unmanned Aviation Operator	720	Yes			
35L	Counterintelligence Agent	1,045	Yes			
35M	Human Intelligence Collector	2,124		Yes		
35N	Signals Intelligence Analyst	1,600		Yes		
35P	Cryptologic Linguist	1,650		Yes		
35S	Signals Analyst	637		Yes		
35T	Military Intelligence System Maintainer	836	Yes			
37F	Psychological Operations Specialist	603		Yes		
38B	Civil Affairs Specialist	171		Yes		
42A	Human Resources Specialist	6,924	Yes		Yes	P, 75TH Ranger
43M	Metal Worker	2	Yes			
45B	Small Arms Repairer	521	Yes		Yes	P, Special Operations Command
45G	Fire Control Repairer	254	Yes			
46Q	Public Affairs Specialist	232	Yes			
46R	Public Affairs Specialist	154	Yes			
56M	Chaplain Assistant	913	Yes		Yes	G/V, 75TH Ranger
63B	Wheeled Vehicle Mechanic	13,167	Yes			
63J	QM and Chemical Equipment Repairer	1,001	Yes			
68K	Medical Lab Specialist	1,057	Yes			

Table B.2—Continued

MOS	MOS Title	Number in Zones A, B, and C	Critical Skill	Special Critical Skill	Location Critical Skill	SQI, Unit
68S	Preventive Medicine Specialist	366	Yes			
68T	Animal Care Specialist	284	Yes			
68W	Health Care Specialist	12,693	Yes		Yes	P, Special Operations Command
68Y		1			Yes	G/V, 75TH Ranger
74D	CBRN Specialist	4,882	Yes		Yes	G/V, 75TH Ranger
79R	Recruiter	769	Yes			
88M	Motor Transportation Operator	10,872	Yes			
88N	Transportation Management Coordinator	1,472	Yes			
89B	Ammunition Specialist	1,971	Yes			
89D	Explosive Ordnance Disposal Specialist	951		Y		
92A	Automated Logistics Specialist	7,495	Yes			
92F	Petrol Supply Specialist	7,336	Yes			
92G	Food Service Specialist	6,205	Yes		Yes	G/V, 75TH Ranger
92R	Parachute Rigger	1,007	Yes			
92W	Water Treatment Specialist	1,603	Yes			
92Y	Unit Supply Specialist	7,771	Yes		Yes	G, V, P; 75TH Ranger and Special Operations Command
94A	Land Missile Repairer	298	Yes			
94E	Comsec Repairer	786	Yes			
94F	Computer System Repairer	824	Yes			
94H	TMDE Specialist	110	Yes			
94M	Radar Repairer	232	Yes			
94S	Patriot System Repairer	77	Yes			
94T	Avenger System Repairer	72	Yes			
94Y	IFTE Operator/Maintainer	134	Yes			
Total Number Eligible For ESRB		216,274				
Total Number Not Eligible		49,934				
Percentage Eligible for ESRB		81.2				

SOURCE: Milpers Message Number 07-344, Army Human Resources Command, December 12, 2007.
NOTE: G refers to Ranger, V refers to Airborne Ranger, and P refers to Parachute Qualified.

Table B.3
Army Enhanced SRB Amounts, March 2007–December 2007

Rank	Months of Additional Obligated Service				
	12–23	24–35	36–48	48–59	60+
Critical Skills List Amounts					
Zone A					
E-3	$6,500	$10,000	$13,500	$16,500	$20,000
E-4	$7,500	$11,000	$14,500	$18,000	$21,500
E-5	$8,000	$12,000	$16,000	$20,000	$23,500
E-6/E-7	$9,000	$13,500	$18,000	$22,000	$26,500
Zone B					
E-4	$8,000	$11,500	$15,500	$19,500	$23,000
E-5	$9,500	$14,000	$19,000	$23,000	$28,000
E-6/E-7	$10,500	$15,500	$20,500	$25,500	$31,000
Zone C					
E-4	$8,000	$11,500	$15,500	$19,500	$23,000
E-5	$10,000	$15,000	$20,000	$24,500	$29,500
E-6/E-7	$11,000	$17,000	$22,500	$28,000	$33,500
Special Critical Skills List Amounts					
Zone A					
E-3	$7,000	$10,000	$13,500	$16,500	$20,000
E-4	$10,000	$13,000	$17,000	$19,500	$23,000
E-5	$15,000	$20,000	$24,000	$26,500	$30,000
E-6/E-7	$20,000	$28,400	$32,400	$34,900	$38,400
Zone B					
E-4	$12,000	$14,000	$18,500	$21,000	$25,000
E-5	$17,000	$22,500	$26,000	$28,500	$32,000
E-6/E-7	$21,000	$29,000	$33,000	$35,500	$39,000
Zone C					
E-4	$12,000	$14,000	$18,500	$21,000	$25,000
E-5	$17,500	$23,000	$27,000	$29,000	$35,000
E-6/E-7	$21,000	$29,000	$34,000	$36,000	$40,000

Table B.4
Army Enhanced SRB Amounts, March 2008–December 2008

Rank	Months of Additional Obligated Service				
	12–23	24–35	36–48	48–59	60+
Critical Skills List Amounts					
Zone A					
E-3	$3,000	$6,000	$9,000	$11,500	$14,500
E-4	$4,000	$7,500	$9,500	$13,000	$15,500
E-5	$5,000	$8,000	$10,000	$15,000	$17,000
E-6/E-7	$6,000	$10,000	$15,000	$17,000	$19,000
Zone B					
E-4	$5,500	$9,000	$12,000	$14,500	$16,500
E-5	$8,000	$10,000	$13,500	$16,500	$20,000
E-6/E-7	$9,500	$11,000	$15,500	$18,500	$22,000
Zone C					
E-4	$4,000	$7,500	$9,500	$13,000	$15,500
E-5	$5,000	$8,000	$10,000	$15,000	$17,000
E-6/E-7	$6,000	$10,000	$15,000	$17,000	$19,000
Special Critical Skills Bonus					
Zone A					
E-3	$5,000	$8,000	$10,000	$12,500	$16,000
E-4	$8,000	$10,000	$12,500	$16,000	$18,000
E-5	$9,500	$11,000	$13,500	$18,000	$20,000
E-6/E-7	$11,000	$13,500	$17,500	$20,000	$24,000
Zone B					
E-4	$8,500	$10,000	$13,000	$17,500	$19,000
E-5	$10,000	$12,000	$14,000	$18,500	$21,000
E-6/E-7	$12,000	$14,000	$18,500	$21,000	$25,000
Zone C					
E-4	$10,000	$12,000	$14,000	$18,500	$21,000
E-5	$12,000	$14,000	$18,500	$21,000	$25,000
E-6/E-7	$15,000	$17,500	$23,000	$27,000	$29,000

SOURCE: Milpers Message Number 08–068, Army Human Resources Command, March 13, 2008.

Table B.5
Army Deployed SRB Amounts, December 2007

Rank	Months of Additional Obligated Service					
	6–11	12–23	24–35	36–48	48–59	60+
Zone A						
E-3	$1,500	$4,500	$6,500	$8,500	$10,000	$13,000
E-4	$2,000	$5,000	$8,000	$10,000	$12,500	$14,000
E-5	$3,000	$6,000	$9,500	$11,000	$13,000	$14,500
E-6/E-7	$4,000	$8,000	$10,000	$12,000	$14,000	$15,000
Zone B						
E-4	$2,000	$5,000	$8,000	$10,000	$12,500	$14,000
E-5	$3,000	$6,000	$9,500	$11,000	$13,000	$14,500
E-6/E-7	$4,000	$8,000	$10,000	$12,000	$14,000	$15,000
Zone C						
E-4	$2,000	$3,500	$5,000	$7,000	$8,000	$9,000
E-5	$2,500	$4,000	$5,500	$7,500	$9,000	$11,000
E-6/E-7	$3,500	$4,500	$6,500	$8,500	$10,000	$13,000

Table B.6
Army Deployed SRB Amounts, December 2008

Rank	Months of Additional Obligated Service					
	6–11	12–23	24–35	36–48	48–59	60+
Zone A						
E-3	$1,500	$4,500	$6,000	$7,500	$9,000	$10,000
E-4	$2,000	$5,000	$6,500	$8,000	$9,500	$11,000
E-5	$3,000	$5,500	$7,000	$8,500	$10,000	$11,500
E-6/E-7	$4,000	$6,000	$7,500	$9,000	$10,500	$12,000
Zone B						
E-4	$2,000	$4,000	$6,500	$8,000	$9,500	$11,000
E-5	$3,000	$5,500	$7,000	$8,500	$10,000	$11,500
E-6/E-7	$4,000	$6,000	$7,500	$9,000	$10,500	$12,000
Zone C						
E-4	$1,500	$3,000	$4,000	$5,000	$6,000	$7,000
E-5	$2,000	$3,500	$5,000	$7,000	$8,000	$9,000
E-6/E-7	$2,500	$4,000	$5,500	$7,500	$9,000	$11,000

Estimated Reenlistment Models, Army 24-MOS Sample

Table C.1
Reenlistment Estimates Using Annual Data from the Army 24-MOS Sample, with SRB Multiplier Conditioned on Deployment Status

Variable	Zones A and B		Zone A		Zone B	
	Estimate	Standard Error	Estimate	Standard Error	Estimate	Standard Error
SRB multiplier	0.056*	0.012	0.059*	0.014	0.070*	0.01
Deployment/Stop-Loss Category (Reference = Not Deployed, No Stop-Loss)						
Deployed, no stop-loss	0.277*	0.018	0.270*	0.020	0.277*	0.01
Deployed, first stop-loss	−0.034	0.041	−0.053	0.042	0.019	0.04
Deployed, continued stop-loss	0.193*	0.049	0.181	0.055	0.210*	0.03
Not deployed, stop-loss	−0.220*	0.016	−0.204*	0.018	−0.240*	0.02
Past Deployment Interval (Reference = 0 Past Deployment)						
1–12 months past deployment	−0.105*	0.023	−0.103*	0.024	−0.098*	0.02
13–24 months past deployment	−0.170*	0.029	−0.164*	0.030	−0.174*	0.03
>24 months past deploment	−0.164*	0.056	−0.115	0.077	−0.279*	0.08
Rank (Reference = E-4)						
Rank E-5	0.403*	0.007	0.427*	0.007	0.307*	0.01
Rank E-6	0.536*	0.008	0.546*	0.009	0.522*	0.01
Years of Service (Reference = YOS 2)						
YOS 3	−0.602*	0.007	−0.599*	0.008		
YOS 4	−0.953*	0.003	−0.984*	0.002		
YOS 5	−0.974*	0.002	−0.994*	0.001		
YOS 6	−0.804*	0.005	−0.856*	0.005		
YOS 7	−0.721*	0.006				
YOS 8	−0.677*	0.007			−0.089*	0.01
YOS 9	−0.621*	0.007			−0.073*	0.01
YOS 10	−0.600*	0.008			−0.062*	0.01
Age	−0.015*	0.001	−0.021*	0.001	−0.001	0.00
Male	0.075*	0.009	0.069*	0.011	0.093*	0.01
High quality	−0.042*	0.006	−0.028*	0.007	−0.065*	0.01

Table C.1—Continued

Variable	Zones A and B		Zone A		Zone B	
	Estimate	Standard Error	Estimate	Standard Error	Estimate	Standard Error
SRB multiplier	0.056*	0.012	0.059*	0.014	0.070*	0.01
Racial-Ethnic Group (Reference = White)						
Black	0.099*	0.006	0.105*	0.007	0.070*	0.01
Hispanic	0.020*	0.005	0.024*	0.006	−0.001	0.01
Other race	0.043*	0.007	0.041*	0.009	0.044*	0.01
Dependents Group (Reference = Unknown Dependents)						
Known dependents	0.066*	0.016	0.061*	0.017	0.127*	0.02
One dependent	0.169*	0.009	0.194*	0.010	0.065*	0.01
Two dependents	0.249*	0.011	0.287*	0.013	0.128*	0.01
Three dependents	0.291*	0.013	0.340*	0.014	0.165*	0.01
Four dependents	0.341*	0.015	0.387*	0.017	0.214*	0.01
Five or more dependents	0.357*	0.013	0.431*	0.015	0.202*	0.01
MOS (Reference = MOS 11B)						
MOS 11C	0.024*	0.011	0.016	0.011	0.063*	0.02
MOS 13B	0.103*	0.013	0.091*	0.015	0.143*	0.02
MOS 13F	0.011	0.014	0.018	0.015	−0.009	0.03
MOS 14S	0.089*	0.020	0.065*	0.024	0.130*	0.02
MOS 14T	0.064	0.037	0.046	0.049	0.123*	0.03
MOS 15T	0.221*	0.029	0.419*	0.025	0.075*	0.03
MOS 15U	0.291*	0.050	0.429*	0.045	0.122*	0.05
MOS 19D	0.033*	0.016	0.027	0.018	0.044*	0.02
MOS 19K	0.124*	0.018	0.116*	0.021	0.146*	0.02
MOS 21B	0.041	0.023	0.032	0.025	0.122*	0.03
MOS 25Q	0.031	0.025	0.050	0.026	−0.030	0.04
MOS 25U	0.149*	0.017	0.176*	0.019	0.077*	0.02
MOS 31B	0.167*	0.023	0.192*	0.028	0.109*	0.03
MOS 35F	0.186*	0.043	0.192*	0.046	0.157	0.10
MOS 35P	0.043	0.033	0.099*	0.039	−0.083*	0.04
MOS 63B	0.182*	0.012	0.177*	0.014	0.181*	0.01
MOS 63H	0.157*	0.018	0.149*	0.022	0.174*	0.02
MOS 63M	0.037	0.019	0.057*	0.024	−0.010	0.04

Table C.1—Continued

Variable	Zones A and B		Zone A		Zone B	
	Estimate	Standard Error	Estimate	Standard Error	Estimate	Standard Error
SRB multiplier	0.056*	0.012	0.059*	0.014	0.070*	0.01
MOS 88M	0.214*	0.016	0.234*	0.019	0.171*	0.02
MOS 92A	0.254*	0.014	0.279*	0.017	0.211*	0.02
MOS 92G	0.206*	0.012	0.187*	0.016	0.219*	0.01
MOS 92F	0.134*	0.014	0.125*	0.015	0.198*	0.02
MOS 92Y	0.163*	0.012	0.157*	0.014	0.178*	0.01
Military Base (Reference = Non-U.S. Base)						
Ft Benning	−0.027	0.044	−0.017	0.049	−0.054	0.03
Ft Bliss	−0.014	0.055	−0.016	0.061	−0.035	0.04
Ft Bragg	−0.070	0.038	−0.073	0.041	−0.041	0.03
Ft Campbell	−0.047	0.050	−0.043	0.052	−0.054	0.04
Ft Carson	−0.079	0.047	−0.055	0.051	−0.150*	0.03
Ft Drum	−0.108*	0.042	−0.097*	0.047	−0.141*	0.04
Ft Hood	−0.080	0.042	−0.067	0.046	−0.112*	0.03
Ft Irwin	−0.105	0.054	−0.070	0.054	−0.200*	0.06
Ft Jackson	0.068	0.039	0.124*	0.055	−0.027	0.05
Ft Knox	−0.070	0.046	−0.012	0.046	−0.173*	0.06
Ft Leavenworth	0.056	0.046	0.097	0.062	−0.054	0.03
Ft Lewis	−0.070	0.044	−0.041	0.048	−0.149*	0.04
Ft Polk	−0.114*	0.042	−0.098*	0.048	−0.165*	0.03
Ft Riley	−0.142*	0.042	−0.127*	0.046	−0.181*	0.03
Ft Sill	−0.086	0.051	−0.085	0.054	−0.094	0.05
Ft Stewart	−0.129*	0.060	−0.108	0.064	−0.196*	0.05
US Alaska	0.012	0.054	0.027	0.053	−0.038	0.07
US Georgia	−0.079*	0.035	−0.036	0.042	−0.180*	0.02
US Hawaii	−0.020	0.047	0.008	0.053	−0.115*	0.04
US Other	0.006	0.039	0.086	0.044	−0.144*	0.02
Unknown	0.068	0.051	0.105	0.053	−0.017	0.05
Reenlistment Decision Year (Reference = FY 2002)						
FY 2003	−0.057*	0.025	−0.054	0.028	−0.059*	0.02
FY 2004	−0.165*	0.025	−0.172*	0.028	−0.126*	0.02

Table C.1—Continued

Variable	Zones A and B		Zone A		Zone B	
	Estimate	Standard Error	Estimate	Standard Error	Estimate	Standard Error
SRB multiplier	0.056*	0.012	0.059*	0.014	0.070*	0.01
FY 2005	−0.178*	0.022	−0.193*	0.023	−0.142*	0.03
FY 2006	−0.213*	0.029	−0.229*	0.030	−0.204*	0.03
No. of observations	119,956		91,468		28,488	
Mean probability		0.489	0.456		0.596	

* Denotes statistical significance at the 5 percent level.

Table C.2
Reenlistment Effects Based on Monthly Data from the Army 24-MOS Sample, with SRB Multiplier Conditioned on Deployment Status, FY 2002–2006

Variable	Zones A and B Marginal Effect	Standard Error	Zone A Marginal Effect	Standard Error	Zone B Marginal Effect	Standard Error
SRB multiplier	0.0050*	0.0011	0.0047*	0.0011	0.0046*	0.0019
Months to ETS Interval (Reference = 0–6 months)						
ETS >1 year	−0.0190	0.0013	−0.017*	0.0011	−0.026*	0.0023
ETS 7–12 months	0.0110*	0.0010	0.0088*	0.0009	0.0171*	0.0020
Deployment/Stop-Loss Category (Reference = Not Deployed, No Stop-Loss)						
Deployed, no stop-loss	0.0017	0.0014	0.0012*	0.0013	0.0077*	0.0029
Not deployed, stop-loss	−0.0143*	0.0010	−0.014*	0.0010	−0.019*	0.0020
Deployed, stop-loss	−0.0123*	0.0012	−0.011*	0.0011	−0.013*	0.0025
Stop-loss months	−0.0005*	0.0002	−0.001*	0.0001	0.0003*	0.0001
Past Deployment Interval (Reference = 0 Past Deployment)						
1–12 months past deployment	−0.0016	0.0015	−0.0017	0.0014	−0.0017	0.0024
13–24 months past deployment	−0.0011	0.0016	−0.0008	0.0015	−0.0020	0.0027
>24 months past deployment	−0.0034	0.0042	−0.0006	0.0068	−0.011*	0.0044
Rank (Reference = E-4)						
Rank E-5	0.0213*	0.0009	0.0187*	0.0009	0.0319*	0.0020
Rank E-6	0.0584*	0.0023	0.0550*	0.0039	0.0641*	0.0030
Rank E-7+	0.1083*	0.0096	0.0011*	0.0009	0.1423*	0.0119
Year of Service Category (Reference = YOS 3)						
YOS 4	0.0004*	0.0010	0.0006	0.0012		
YOS 5	−0.0008	0.0012	−0.005*	0.0009		
YOS 6	−0.0061*	0.0010				
YOS 7	−0.0070	0.0009				
YOS 8	−0.0048*	0.0010			0.0029*	0.0014
YOS 9	−0.0046*	0.0010			0.0034*	0.0014
YOS 10	−0.0015	0.0013			0.0096*	0.0019
Age	−0.0007*	0.0001	−0.001*	0.0001	−0.0004	0.0002
High quality	−0.0079*	0.0005	−0.009*	0.0006	−0.004*	0.0011
Racial-Ethnic Group (Reference = White)						
Black	0.0052*	0.0008	0.0049*	0.0009	0.0062*	0.0017
Hispanic	−0.0010	0.0007	−0.0008	0.0007	−0.0019	0.0017

Table C.2—Continued

Variable	Zones A and B		Zone A		Zone B	
	Marginal Effect	Standard Error	Marginal Effect	Standard Error	Marginal Effect	Standard Error
SRB multiplier	0.0050*	0.0011	0.0047*	0.0011	0.0046*	0.0019
Other race	0.0008	0.0009	0.0008	0.0010	−0.0002	0.0025
Dependents Group (Reference = Unknown Dependents)						
Known dependents	−0.0036*	0.0015	−0.003*	0.0015	−0.0003	0.0024
One dependent	0.0121*	0.0007	0.0116*	0.0007	0.0084*	0.0016
Two dependents	0.0210*	0.0012	0.0216*	0.0013	0.0160*	0.0019
Three dependents	0.0259*	0.0015	0.0263*	0.0020	0.0228*	0.0024
Four dependents	0.0328*	0.0029	0.0383*	0.0044	0.0257*	0.0034
> Five dependents	0.0392*	0.0036	0.0510*	0.0068	0.0313*	0.0043
Military Base (Reference = Non-U.S. Base)						
Ft Benning	−0.0078*	0.0016	−0.005*	0.0016	−0.017*	0.0023
Ft Bliss	−0.0013	0.0022	−0.0006	0.0023	−0.0031	0.0028
Ft Bragg	−0.0027	0.0016	−0.0009	0.0017	−0.0049	0.0026
Ft Campbell	−0.0077*	0.0015	−0.005*	0.0016	−0.011*	0.0034
Ft Carson	0.0001	0.0016	0.0004	0.0015	0.0008	0.0044
Ft Drum	0.0181	0.0096	0.0308*	0.0072	0.0124	0.0120
Ft Hood	−0.0024	0.0021	−0.0026	0.0020	−0.0001	0.0038
Ft Irwin	−0.0029	0.0018	−0.0023	0.0020	−0.0041	0.0042
Ft Jackson	−0.0017	0.0017	−0.0009	0.0018	−0.0045	0.0037
Ft Knox	−0.0057*	0.0011	−0.0048	0.0012	−0.008*	0.0034
Ft Leavenworth	−0.0006	0.0017	0.0006	0.0017	−0.0047	0.0035
Ft Lewis	−0.0016	0.0033	0.0009	0.0033	−0.0096	0.0047
Ft Polk	−0.0007	0.0031	0.0005	0.0071	−0.0007	0.0048
Ft Riley	0.0031	0.0098	−0.0024	0.0086	0.0096	0.0136
Ft Sill	0.0018	0.0021	0.0031	0.0022	−0.002*	0.0051
Ft Stewart	−0.0040*	0.0016	−0.0025	0.0016	−0.009*	0.0027
US Alaska	−0.0022	0.0023	−0.0031	0.0019	0.0018	0.0055
US Georgia	−0.0056	0.0031	−0.0040	0.0027	−0.0108	0.0051
US Hawaii	0.0014	0.0025	0.0000	0.0025	0.0050*	0.0037
US Other	−0.0057	0.0013	−0.005*	0.0015	−0.008*	0.0036
Unknown	0.0000	0.0023	−0.0006	0.0021	0.0027*	0.0041

Table C.2—Continued

Variable	Zones A and B		Zone A		Zone B	
	Marginal Effect	Standard Error	Marginal Effect	Standard Error	Marginal Effect	Standard Error
Fiscal Year (Reference = FY 2002)						
FY 2003	−0.0037	0.0021	−0.0026	0.0026	−0.008*	0.0030
FY 2004	−0.0019	0.0015	−0.0008	0.0017	−0.008*	0.0026
FY 2005	−0.0027	0.0014	−0.0014	0.0016	−0.007*	0.0028
FY 2006	0.0018	0.0018	0.0042*	0.0016	−0.0022	0.0042
MOS (Reference = MOS 11B)						
MOS 13B	−0.0048*	0.0007	−0.0049	0.0008	−0.004*	0.0019
MOS 13F	−0.0101*	0.0007	−0.0081	0.0008	−0.015*	0.0019
MOS 21B	−0.0067*	0.0007	−0.0059	0.0008	−0.009*	0.0017
No. of observations	558,312		425,178		129,849	
Mean probability	0.031		0.026		0.047	

* Denotes statistical significance at the 5 percent level.

Table C.3
Tobit Regressions for Months of Reenlistment

| | FY 2002–2006 | | | | FY 2002–2004 | | | |
| | Zone A | | Zone B | | Zone A | | Zone B | |
	Parameter Estimate	Standard Error	Parameter Estimate	Standard Error	Parameter Estimate	Standard Error	Parameter Estimate	Standard Error
SRB Multiplier (Reference Group = 0)								
SRBM = 0.5	4.55*	0.50	10.52*	0.90	4.65*	0.56	10.19*	0.99
SRBM = 1.0	8.14*	0.46	13.33*	0.46	6.78*	0.46	12.50*	0.55
SRBM = 1.5	9.56*	0.49	11.58*	0.64	9.57*	0.65	14.77*	1.03
SRBM = 2.0	11.06*	0.65	11.07*	0.93	13.95*	0.61	18.27*	1.04
SRBM = 2.5	7.93*	1.60	5.88*	0.74	15.63*	0.73	10.04*	2.58
SRBM = 3.0	4.01*	0.65	3.73*	0.77	10.84*	1.24	8.57*	1.70
SRBM > 3.0	1.58	0.99	1.09	0.99	10.08*	1.46	20.11*	2.34
Deployment/Stop-Loss Category (Reference = Not Deployed, No Stop-Loss)								
Deployed, no stop-loss	3.30*	0.43	3.18*	0.54	1.60*	0.60	3.00*	0.88
Deployed, stop-loss	6.86*	0.83	8.42*	0.85	8.56*	0.77	11.59*	1.29
Not deployed, stop-loss	2.11*	0.50	2.17*	0.90	2.90*	0.54	2.31	1.47
1–12 months past deployment	–1.67*	0.43	–1.82*	0.45	–0.88	0.57	–1.65*	0.61
13–24 months past deployment	–2.27*	0.40	–2.41*	0.58	–1.73	1.41	–4.13	2.62
>24 months past deployment	–2.44	1.33	–3.47	3.18	–0.27	0.19		
Rank (Reference = E-4)								
Rank E-5	–0.30	0.16	2.79*	0.40	1.17	0.71	2.83*	0.53
Rank E-6	0.04	0.47	3.36*	0.45	2.86	2.12	3.62*	0.59
Year of Service Category (Reference = YOS 2 [Zone A] or YOS 7 [Zone B])								
YOS 3	2.44	2.12			2.40	2.10		
YOS 4	2.06	2.10			2.12	2.10		
YOS 5	1.81	2.09			0.84	2.04		
YOS 6	0.39	2.06						
YOS 8			2.17*	0.26			1.94*	0.30
YOS 9			2.67*	0.31			2.63*	0.38
YOS 10			4.25*	0.38			4.78*	0.43
Age	0.08*	0.02	0.10*	0.04	0.10*	0.02	0.12*	0.04
Male	1.07*	0.19	0.96*	0.26	1.16*	0.22	0.91*	0.35

Table C.3—Continued

	FY 2002–2006				FY 2002–2004			
	Zone A		Zone B		Zone A		Zone B	
	Parameter Estimate	Standard Error	Parameter Estimate	Standard Error	Parameter Estimate	Standard Error	Parameter Estimate	Standard Error
Racial-Ethnic Group (Reference = White)								
Black	−0.78*	0.16	−1.16*	0.21	−0.83*	0.17	−1.03*	0.25
Hispanic	−1.49*	0.18	−0.72*	0.30	−1.44*	0.24	−0.28	0.39
Other race	−0.77*	0.25	−0.74	0.40	−0.87*	0.31	−0.93	0.55
High quality	−0.77*	0.12	0.30	0.21	−0.53*	0.15	0.52*	0.24
Dependents Group (Reference = Unknown Dependents)								
Known dependents	−0.67*	0.30	−1.28*	0.41	−0.55	0.30	−1.37*	0.47
One dependent	0.83*	0.17	1.31*	0.27	0.94*	0.21	1.38*	0.34
Two dependents	1.25*	0.18	1.40*	0.33	1.26*	0.21	1.35*	0.45
Three dependents	2.02*	0.25	1.87*	0.36	1.86*	0.30	1.95*	0.46
Four dependents	2.51*	0.37	1.62*	0.41	3.25*	0.44	2.18*	0.53
Five or more dependents	2.57*	0.64	2.33*	0.59	2.63*	0.91	2.15*	0.80
MOS (Reference = MOS 11B) and Home Base (Reference = Europe)								
MOS 11C	−0.72	0.42	1.31	0.86	−0.42	0.55	−0.85	0.95
MOS 13B	−0.53	0.42	−0.05	0.59	−0.29	0.44	−1.02	0.79
MOS 13F	0.44	0.53	1.29	0.69	1.58*	0.62	1.15	0.90
MOS 14S	−0.63	0.72	0.48	1.00	−1.26	0.64	−1.72	1.27
MOS 14T	3.58*	1.10	1.76	1.20	2.20*	0.85	−0.66	1.42
MOS 15T	3.06*	0.58	3.66*	1.04	4.65*	0.62	2.88*	1.10
MOS 15U	3.99*	1.02	5.53*	1.62	3.31*	1.05	1.44	1.88
MOS 19D	0.08	0.43	2.66*	0.58	0.83	0.50	2.28*	0.82
MOS 19K	0.59	0.48	1.67*	0.64	0.58	0.60	0.04	0.76
MOS 21B	−1.15*	0.41	−2.26*	0.87	−1.68*	0.47	−3.37*	0.88
MOS 25Q	0.87	0.52	3.93*	0.98	−1.47*	0.47	1.55	1.01
MOS 25U	−1.37*	0.39	−1.45*	0.67	−1.63*	0.45	−2.58*	0.69
MOS 31B	−3.78*	0.43	3.13*	0.80	−1.91*	0.48	1.52	0.92
MOS 35F	−1.82	1.87	−1.69	1.56	−1.90	1.84	−2.94	1.47
MOS 35P	1.13	1.15	0.25	1.32	−2.22*	0.95	−1.31	1.86
MOS 63B	−1.34*	0.34	−0.46	0.45	−1.47*	0.33	−1.88*	0.58
MOS 63H	−0.24	0.58	−0.59	0.83	−0.39	0.73	−3.10*	1.14

Table C.3—Continued

| | FY 2002–2006 | | | | FY 2002–2004 | | | |
| | Zone A | | Zone B | | Zone A | | Zone B | |
	Parameter Estimate	Standard Error	Parameter Estimate	Standard Error	Parameter Estimate	Standard Error	Parameter Estimate	Standard Error
MOS 63M	2.07*	0.73	−1.90	1.53	2.99*	0.82	−3.22*	1.59
MOS 88M	−2.26*	0.45	0.12	0.54	−1.69*	0.47	−1.09	0.75
MOS 92A	−1.31*	0.40	−0.84	0.57	−2.01*	0.35	−2.71*	0.66
MOS 92G	0.36	0.48	1.68*	0.66	0.69	0.60	−0.68	0.79
MOS 92F	−0.66	0.39	−3.58*	0.61	−0.64	0.50	−4.57*	0.68
MOS 92Y	−2.20*	0.36	−1.97*	0.57	−2.36*	0.40	−3.31*	0.55
Ft Benning	1.49*	0.61	−0.05	0.73	1.62*	0.67	−1.47*	0.50
Ft Bliss	0.03	0.72	−0.77	1.07	0.04	1.25	0.20	1.72
Ft Bragg	0.11	0.54	0.22	0.62	0.45	0.51	−0.07	0.77
Ft Campbell	0.11	0.61	0.52	0.41	0.40	0.41	0.44	0.59
Ft Carson	0.16	0.77	0.37	1.05	0.47	0.97	0.95	1.40
Ft Drum	0.13	0.38	−0.67	0.78	−0.11	0.41	−0.91	1.07
Ft Hood	1.44*	0.71	1.17	0.70	2.02*	0.82	1.69*	0.57
Ft Irwin	−3.71*	0.80	−2.69*	1.04	−3.75*	1.05	−2.37	1.34
Ft Jackson	−1.83	1.49	−1.18	0.87	−2.46	1.92	−1.57	1.26
Ft Knox	−2.14*	0.73	−1.45	1.08	−2.17*	1.07	−2.58*	0.80
Ft Leavenworth	−0.19	0.75	0.97	1.18	0.23	0.80	1.90	1.18
Ft Lewis	−0.54	0.51	−0.52	0.74	−0.23	0.49	−0.74	1.07
Ft Polk	2.03*	0.82	0.37	0.46	2.53*	0.91	0.50	0.55
Ft Riley	0.09	0.80	−0.52	0.74	0.55	0.89	−0.65	0.45
Ft Sill	−0.31	1.10	−1.85*	0.73	−0.93	1.28	−1.62*	0.49
Ft Stewart	2.19*	0.80	1.72	1.18	2.20*	0.85	1.54	1.81
US Alaska	0.49	0.52	−0.32	0.48	0.10	0.41	−0.73	0.53
US Georgia	0.43	0.56	1.10	0.77	0.49	0.69	0.64	0.91
US Hawaii	1.17*	0.58	1.41	1.24	1.21	0.85	1.25	1.78
US Other	−1.09	0.56	−1.01	0.67	−0.65	0.86	−1.29	0.91
Unknown	−0.68	0.43	−0.05	0.58	0.02	0.40	−0.63	0.49
FY 2003	−2.13*	0.32	−2.88*	0.46	−1.79*	0.33	−2.81*	0.51
FY 2004	1.87*	0.53	3.19*	0.60	2.37*	0.51	3.05*	0.58
FY 2005	1.20*	0.42	2.46*	0.56				

Table C.3—Continued

| | FY 2002–2006 | | | | FY 2002–2004 | | | |
| | Zone A | | Zone B | | Zone A | | Zone B | |
	Parameter Estimate	Standard Error	Parameter Estimate	Standard Error	Parameter Estimate	Standard Error	Parameter Estimate	Standard Error
FY 2006	0.87	0.74	1.17*	0.58				
Intercept	31.58*	2.27	28.24	1.14	29.99	2.38	28.50	1.42
Standard error	10.87		12.26		10.80		12.33	
No. of observations	40,730		15,719		27,960		10,186	

* Denotes statistical significance at the 5 percent level.

Estimated Reenlistment Models, All Services

Table D.1
Estimates from the Reenlistment Model, Army

	SRBM Varies by Deployment?			
	First Term		Second Term	
	No	Yes	No	Yes
SRB multiplier	0.025**	0.089**	0.025**	0.051**
	(0.002)	(0.002)	(0.002)	(0.002)
Any hostile fire pay deployment	0.087**	0.087**	0.075**	0.071**
	(0.007)	(0.007)	(0.005)	(0.005)
Only non–hostile fire pay deployment	−0.009**	−0.012**	0.015**	0.014**
	(0.003)	(0.003)	(0.003)	(0.003)
Years of service	0.013**	0.018**	−0.037**	−0.032**
	(0.002)	(0.002)	(0.001)	(0.001)
High school dropout or education missing	0.012	0.013	0.008	0.005
	(0.010)	(0.010)	(0.011)	(0.011)
General Equivalency Diploma	0.055**	0.052**	0.027**	0.022**
	(0.007)	(0.007)	(0.008)	(0.008)
At least some college	−0.100**	−0.096**	−0.016*	−0.013†
	(0.006)	(0.006)	(0.007)	(0.007)
Male	0.032**	0.030**	0.040**	0.036**
	(0.004)	(0.004)	(0.004)	(0.004)
AFQT less than Cat IIIB or missing	−0.030**	−0.031**	0.066**	0.060**
	(0.011)	(0.011)	(0.008)	(0.008)
AFQT Cat IIIB	−0.099**	−0.098**	0.015**	0.016**
	(0.003)	(0.003)	(0.004)	(0.004)
AFQT Cat II	−0.029**	−0.030**	−0.020**	−0.017**
	(0.003)	(0.003)	(0.004)	(0.004)
AFQT Cat I	−0.069**	−0.068**	−0.064**	−0.060**
	(0.006)	(0.006)	(0.009)	(0.009)
White	−0.006†	−0.007*	0.007	0.006
	(0.003)	(0.003)	(0.004)	(0.004)
Black	0.084**	0.081**	0.053**	0.051**
	(0.004)	(0.004)	(0.004)	(0.004)
Promoted rapidly	0.340**	0.339**	0.119**	0.114**
	(0.003)	(0.003)	(0.004)	(0.004)
No. of observations	129,322	123,774	85,969	84,055

NOTES: Cell entries are ordinary least squares regression coefficients for a linear probability model in which the dependent variable is 1 for reenlistment (additional obligated service of 24 months or more) or 0. Models also include controls for service duty MOS fixed effects and indicators for year of decision (year fixed effects). Robust standard errors are in parentheses.

** Denotes statistical significance at the 1 percent level.

* Denotes statistical significance at the 5 percent level.

† Denotes statistical significance at the 10 percent level.

Table D.2
Estimates from the Reenlistment Model, Navy

| | SRBM Varies by Deployment? | | | |
| | First Term | | Second Term | |
	No	Yes	No	Yes
SRB multiplier	0.025**	0.025**	0.010**	0.009**
	(0.002)	(0.002)	(0.002)	(0.001)
Any hostile fire pay deployment	0.093**	0.093**	0.117**	0.117**
	(0.005)	(0.005)	(0.005)	(0.005)
Only non–hostile fire pay deployment	−0.001	0.002	0.041**	0.049**
	(0.003)	(0.003)	(0.004)	(0.004)
Years of service	−0.102**	−0.099**	0.031**	0.029**
	(0.003)	(0.003)	(0.001)	(0.001)
High school dropout or education missing	0.049**	0.048**	0.025†	0.021
	(0.009)	(0.009)	(0.014)	(0.014)
General Equivalency Diploma	0.069**	0.066**	−0.003	−0.003
	(0.012)	(0.012)	(0.012)	(0.012)
At least some college	−0.031**	−0.029**	0.004	0.003
	(0.010)	(0.010)	(0.009)	(0.009)
Male	0.034**	0.036**	0.058**	0.059**
	(0.004)	(0.004)	(0.005)	(0.005)
AFQT less than Cat IIIB or missing	0.098**	0.100**	0.065**	0.066**
	(0.027)	(0.027)	(0.017)	(0.016)
AFQT Cat IIIB	0.061**	0.062**	0.035**	0.033**
	(0.004)	(0.004)	(0.005)	(0.005)
AFQT Cat II	−0.043**	−0.042**	−0.023**	−0.021**
	(0.004)	(0.004)	(0.005)	(0.005)
AFQT Cat I	−0.084**	−0.080**	−0.084**	−0.079**
	(0.010)	(0.010)	(0.010)	(0.010)
White	−0.061**	−0.061**	−0.035**	−0.035**
	(0.004)	(0.004)	(0.005)	(0.005)
Black	0.110**	0.109**	0.042**	0.041**
	(0.004)	(0.004)	(0.005)	(0.005)
Promoted rapidly	0.274**	0.272**	0.127**	0.121**
	(0.003)	(0.003)	(0.004)	(0.004)
No. of observations	96,334	93,949	58,107	57,104

NOTES: Cell entries are ordinary least squares regression coefficients for a linear probability model in which the dependent variable is 1 for reenlistment (additional obligated service of 24 months or more) or 0. Models also include controls for service duty MOS fixed effects and year-of-decision indicators (year fixed effects). Robust standard errors are in parentheses.

** Denotes statistical significance at the 1 percent level.

† Denotes statistical significance at the 10 percent level.

Table D.3
Estimates from the Reenlistment Model, Marine Corps

| | SRBM Varies by Deployment? | | | |
| | First Term | | Second Term | |
	No	Yes	No	Yes
SRB multiplier	0.036** (0.002)	0.035** (0.002)	−0.004 (0.004)	−0.003 (0.004)
Any hostile fire pay deployment	0.096** (0.006)	0.096** (0.006)	0.103** (0.008)	0.101** (0.007)
Only non–hostile fire pay deployment	0.023** (0.004)	0.024** (0.004)	0.058** (0.007)	0.058** (0.007)
Years of service	−0.331** (0.008)	−0.327** (0.008)	−0.015** (0.002)	−0.014** (0.002)
High school dropout or education missing	−0.029 (0.023)	−0.034 (0.023)	−0.060 (0.127)	−0.068 (0.128)
General Equivalency Diploma	0.027 (0.017)	0.030† (0.017)	0.038† (0.021)	0.038† (0.021)
At least some college	−0.011 (0.014)	−0.014 (0.015)	−0.029* (0.015)	−0.028† (0.015)
Male	0.011 (0.007)	0.009 (0.007)	0.093** (0.012)	0.088** (0.012)
AFQT less than Cat IIIB or missing	−0.018 (0.017)	−0.024 (0.018)	0.024 (0.030)	0.016 (0.030)
AFQT Cat IIIB	−0.004 (0.004)	−0.004 (0.004)	0.014† (0.007)	0.013† (0.007)
AFQT Cat II	−0.018** (0.004)	−0.017** (0.004)	−0.027** (0.007)	−0.029** (0.007)
AFQT Cat I	−0.049** (0.009)	−0.053** (0.009)	−0.042* (0.017)	−0.036* (0.017)
White	−0.015** (0.004)	−0.014** (0.004)	−0.013† (0.007)	−0.014* (0.007)
Black	0.104** (0.006)	0.103** (0.006)	0.056** (0.009)	0.055** (0.009)
Promoted rapidly	0.068** (0.005)	0.055** (0.006)	0.148** (0.006)	0.146** (0.006)
No. of observations	77,214	73,744	22,489	22,143

NOTES: Cell entries are ordinary least squares regression coefficients for a linear probability model in which the dependent variable is 1 for reenlistment (additional obligated service of 24 months or more) or 0. Models also include controls for service duty MOS fixed effects and year-of-decision indicators (year fixed effects). Robust standard errors are in parentheses.
** Denotes statistical significance at the 1 percent level.
* Denotes statistical significance at the 5 percent level.
† Denotes statistical significance at the 10 percent level.

Table D.4
Estimates from the Reenlistment Model, Air Force

	SRBM Varies by Deployment?			
	First Term		Second Term	
	No	Yes	No	Yes
SRB multiplier	0.016** (0.002)	0.013** (0.001)	0.015** (0.002)	0.014** (0.002)
Any hostile fire pay deployment	0.148** (0.007)	0.145** (0.007)	0.065** (0.007)	0.063** (0.006)
Only non–hostile fire pay deployment	−0.005 (0.004)	0.001 (0.004)	0.023** (0.005)	0.026** (0.005)
Years of service	−0.036** (0.002)	−0.035** (0.002)	−0.004** (0.001)	−0.004** (0.001)
High school dropout or education missing	−0.044 (0.064)	−0.051 (0.064)	−0.244 (0.172)	−0.262 (0.169)
General Equivalency Diploma	0.098 (0.065)	0.079 (0.065)	0.000	0.000
At least some college	−0.114** (0.006)	−0.112** (0.006)	−0.095** (0.005)	−0.094** (0.005)
Male	0.023** (0.004)	0.024** (0.004)	0.053** (0.005)	0.051** (0.005)
AFQT less than Cat IIIB or missing	0.035 (0.037)	0.038 (0.036)	0.010 (0.036)	0.004 (0.036)
AFQT Cat IIIB	0.016** (0.005)	0.016** (0.005)	0.009† (0.005)	0.008 (0.005)
AFQT Cat II	−0.027** (0.004)	−0.026** (0.004)	−0.002 (0.005)	−0.004 (0.005)
AFQT Cat I	−0.070** (0.008)	−0.066** (0.008)	−0.015 (0.010)	−0.012 (0.010)
White	−0.023** (0.005)	−0.023** (0.005)	−0.025** (0.006)	−0.024** (0.006)
Black	0.065** (0.006)	0.065** (0.006)	0.015* (0.007)	0.015* (0.007)
Promoted rapidly	−0.021** (0.005)	−0.026** (0.005)	0.022* (0.009)	0.012 (0.009)
No. of observations	87,707	86,173	42,470	42,161

NOTES: Cell entries are ordinary least squares regression coefficients for a linear probability model in which the dependent variable is 1 for reenlistment (additional obligated service of 24 months or more) or 0. Models also include controls for service duty MOS fixed effects and year-of-decision indicators (year fixed effects). Robust standard errors are in parentheses.
** Denotes statistical significance at the 1 percent level.
* Denotes statistical significance at the 5 percent level.
† Denotes statistical significance at the 10 percent level.

Table D.5
Estimates from the Length of Reenlistment Model, Army

	SRBM Varies by Deployment?			
	First Term		Second Term	
	No	Yes	No	Yes
SRBM 0.5	−0.565**	−0.648**	1.454**	1.352**
	(0.210)	(0.199)	(0.187)	(0.179)
SRBM 1	2.062**	5.555**	4.425**	5.799**
	(0.244)	(0.237)	(0.195)	(0.188)
SRBM 1.5	4.736**	8.473**	7.004**	9.425**
	(0.253)	(0.233)	(0.202)	(0.188)
SRBM 2	5.096**	13.331**	7.416**	10.820**
	(0.267)	(0.236)	(0.218)	(0.196)
SRBM 2.5	4.823**	11.958**	7.226**	9.496**
	(0.328)	(0.306)	(0.252)	(0.233)
SRBM 3	6.987**	15.963**	6.015**	9.433**
	(0.424)	(0.376)	(0.302)	(0.272)
SRBM >3	5.118**	12.115**	3.006**	4.749**
	(0.480)	(0.431)	(0.317)	(0.285)
Any hostile fire pay deployment	3.120**	2.978**	1.948**	1.693**
	(0.311)	(0.300)	(0.197)	(0.191)
Only non–hostile fire pay deployment	−0.395**	−0.655**	1.035**	0.889**
	(0.145)	(0.141)	(0.119)	(0.115)
Years of service	1.778**	1.928**	0.539**	0.696**
	(0.073)	(0.071)	(0.032)	(0.031)
High school dropout or education missing	0.279	0.252	0.574	0.429
	(0.512)	(0.497)	(0.407)	(0.396)
General Equivalency Diploma	2.821**	2.569**	0.507†	0.394
	(0.363)	(0.353)	(0.307)	(0.297)
At least some college	−4.504**	−4.034**	0.259	0.354
	(0.295)	(0.286)	(0.248)	(0.241)
Male	2.578**	2.382**	2.355**	2.115**
	(0.191)	(0.184)	(0.160)	(0.155)
AFQT less than Cat IIIB or missing	0.192	0.230	2.016**	1.825**
	(0.501)	(0.483)	(0.322)	(0.312)
AFQT Cat IIIB	−4.260**	−3.923**	0.482**	0.522**
	(0.173)	(0.167)	(0.136)	(0.132)
AFQT Cat II	−1.258**	−1.253**	−0.503**	−0.414**
	(0.162)	(0.157)	(0.144)	(0.139)
AFQT Cat I	−3.371**	−3.191**	−1.986**	−1.854**
	(0.326)	(0.316)	(0.331)	(0.322)
White	0.279	0.266	0.738**	0.674**
	(0.177)	(0.171)	(0.148)	(0.144)
Black	3.698**	3.360**	1.542**	1.398**
	(0.205)	(0.198)	(0.159)	(0.154)
Promoted rapidly	16.885**	15.975**	5.752**	5.226**
	(0.169)	(0.164)	(0.124)	(0.121)
No. of observations	129,322	123,774	85,969	84,055

NOTES: Cell entries are estimated Tobit regression coefficients. Models also include controls for service duty MOS fixed effects and year-of-decision indicators. Standard errors are in parentheses.

** Denotes statistical significance at the 1 percent level.

† Denotes statistical significance at the 10 percent level.

Table D.6
Estimates from the Length of Reenlistment Model, Navy

	SRBM Varies by Deployment?			
	First Term		Second Term	
	No	Yes	No	Yes
SRBM 0.5	3.740**	3.373**	5.230**	4.881**
	(0.292)	(0.295)	(0.349)	(0.348)
SRBM 1	5.802**	5.441**	8.855**	8.851**
	(0.268)	(0.262)	(0.337)	(0.331)
SRBM 1.5	8.094**	8.409**	10.940**	10.619**
	(0.275)	(0.271)	(0.358)	(0.351)
SRBM 2	10.021**	9.868**	11.642**	11.725**
	(0.296)	(0.284)	(0.393)	(0.386)
SRBM 2.5	11.536**	11.047**	11.482**	11.293**
	(0.329)	(0.324)	(0.411)	(0.404)
SRBM 3	12.138**	11.710**	10.559**	10.204**
	(0.388)	(0.378)	(0.436)	(0.429)
SRBM >3	11.691**	11.172**	5.824**	5.477**
	(0.330)	(0.318)	(0.345)	(0.337)
Any hostile fire pay deployment	4.319**	4.256**	3.991**	3.943**
	(0.221)	(0.218)	(0.271)	(0.268)
Only non–hostile fire pay deployment	−0.519**	−0.372*	0.326	0.759**
	(0.151)	(0.150)	(0.207)	(0.206)
Years of service	−2.488**	−2.242**	2.218**	2.131**
	(0.121)	(0.120)	(0.048)	(0.047)
High school dropout or education missing	3.131**	3.077**	1.504*	1.333†
	(0.416)	(0.412)	(0.709)	(0.701)
General Equivalency Diploma	2.643**	2.442**	0.620	0.617
	(0.554)	(0.549)	(0.567)	(0.561)
At least some college	−0.958†	−0.889†	−0.532	−0.618
	(0.493)	(0.489)	(0.432)	(0.428)
Male	1.890**	1.950**	3.306**	3.293**
	(0.192)	(0.190)	(0.246)	(0.244)
AFQT less than Cat IIIB or missing	5.087**	5.093**	0.127	0.189
	(1.230)	(1.218)	(0.944)	(0.935)
AFQT Cat IIIB	2.453**	2.392**	1.202**	1.075**
	(0.180)	(0.178)	(0.236)	(0.233)
AFQT Cat II	−2.059**	−1.973**	−1.057**	−0.974**
	(0.192)	(0.190)	(0.236)	(0.234)
AFQT Cat I	−4.329**	−4.090**	−5.263**	−5.013**
	(0.477)	(0.474)	(0.425)	(0.423)
White	−2.280**	−2.185**	−0.427*	−0.399†
	(0.173)	(0.171)	(0.216)	(0.214)
Black	4.447**	4.309**	1.861**	1.808**
	(0.203)	(0.201)	(0.254)	(0.251)
Promoted rapidly	13.471**	13.221**	5.943**	5.716**
	(0.158)	(0.157)	(0.210)	(0.208)
No. of observations	96,334	93,949	58,107	57,104

NOTES: Cell entries are estimated Tobit regression coefficients. Models also include controls for service duty MOS fixed effects and year-of-decision indicators. Standard errors are in parentheses.

** Denotes statistical significance at the 1 percent level.

* Denotes statistical significance at the 5 percent level.

† Denotes statistical significance at the 10 percent level.

Table D.7
Estimates from the Length of Reenlistment Model, Marine Corps

	SRBM Varies by Deployment?			
	First Term		Second Term	
	No	Yes	No	Yes
SRBM 0.5	6.263**	6.741**	2.692**	2.366**
	(0.391)	(0.393)	(0.299)	(0.296)
SRBM 1	8.629**	7.634**	3.266**	3.545**
	(0.303)	(0.295)	(0.287)	(0.281)
SRBM 1.5	6.502**	6.550**	3.867**	4.094**
	(0.346)	(0.336)	(0.353)	(0.349)
SRBM 2	8.879**	9.076**	4.044**	4.202**
	(0.468)	(0.470)	(0.465)	(0.450)
SRBM 2.5	9.199**	8.523**	4.494**	4.486**
	(0.542)	(0.530)	(0.528)	(0.512)
SRBM 3	11.118**	10.830**	3.951**	3.818**
	(0.542)	(0.532)	(0.577)	(0.564)
SRBM >3	12.662**	11.194**	1.281*	0.993†
	(0.386)	(0.370)	(0.538)	(0.524)
Any hostile fire pay deployment	4.722**	4.536**	2.214**	2.111**
	(0.294)	(0.289)	(0.253)	(0.247)
Only non–hostile fire pay deployment	1.317**	1.323**	2.079**	2.107**
	(0.198)	(0.195)	(0.206)	(0.202)
Years of service	–13.786**	–13.080**	0.056	0.103
	(0.172)	(0.168)	(0.069)	(0.067)
High school dropout or education missing	–1.902	–1.962	–3.033	–3.228
	(1.412)	(1.388)	(3.470)	(3.373)
General Equivalency Diploma	1.254	1.382†	0.710	0.717
	(0.838)	(0.824)	(0.622)	(0.609)
At least some college	–0.227	–0.384	–0.591	–0.545
	(0.711)	(0.698)	(0.437)	(0.428)
Male	0.758*	0.634†	2.744**	2.563**
	(0.352)	(0.347)	(0.343)	(0.336)
AFQT less than Cat IIIB or missing	–0.530	–0.840	–0.029	–0.249
	(0.942)	(0.921)	(0.904)	(0.881)
AFQT Cat IIIB	–0.080	–0.086	0.295	0.272
	(0.222)	(0.219)	(0.216)	(0.212)
AFQT Cat II	–0.915**	–0.822**	–0.568**	–0.616**
	(0.222)	(0.219)	(0.219)	(0.215)
AFQT Cat I	–2.528**	–2.608**	–0.507	–0.266
	(0.501)	(0.494)	(0.519)	(0.510)
White	–0.594**	–0.517*	–0.233	–0.253
	(0.218)	(0.214)	(0.214)	(0.210)
Black	4.869**	4.680**	1.154**	1.103**
	(0.304)	(0.299)	(0.264)	(0.258)
Promoted rapidly	4.775**	3.911**	3.775**	3.651**
	(0.304)	(0.302)	(0.176)	(0.172)
No. of observations	77,214	73,744	22,489	22143

NOTES: Cell entries are estimated Tobit regression coefficients. Models also include controls for service duty MOS fixed effects and year-of-decision indicators. Standard errors are in parentheses.

** Denotes statistical significance at the 1 percent level.

* Denotes statistical significance at the 5 percent level.

† Denotes statistical significance at the 10 percent level.

Table D.8
Estimates from the Length of Reenlistment Model, Air Force

| | SRBM Varies by Deployment? | | | |
| | No | Yes | No | Yes |
	First Term		Second Term	
SRBM 0.5	−0.550 (0.357)	−1.027** (0.362)	0.923** (0.335)	0.846** (0.327)
SRBM 1	−1.681** (0.349)	−2.377** (0.350)	2.057** (0.327)	1.960** (0.318)
SRBM 1.5	−2.568** (0.368)	−3.195** (0.364)	2.043** (0.359)	1.729** (0.351)
SRBM 2	2.605** (0.350)	2.187** (0.342)	2.050** (0.400)	1.960** (0.389)
SRBM 2.5	3.949** (0.365)	4.079** (0.358)	1.910** (0.447)	1.654** (0.437)
SRBM 3	3.920** (0.426)	3.438** (0.424)	2.519** (0.506)	1.661** (0.495)
SRBM >3	5.545** (0.318)	5.035** (0.310)	1.072* (0.419)	0.512 (0.406)
Any HFP deployment	6.174** (0.340)	5.935** (0.333)	2.988** (0.322)	2.917** (0.318)
Only non–HFP deployment	−0.538** (0.173)	−0.250 (0.171)	0.534** (0.202)	0.647** (0.201)
Years of service	−1.156** (0.087)	−1.113** (0.085)	1.275** (0.061)	1.261** (0.060)
High school dropout or education missing	−1.850 (3.181)	−2.399 (3.111)	−7.806 (5.871)	−8.471 (5.802)
General Equivalency Diploma	4.129 (3.270)	3.381 (3.199)		
At least some college	−5.561** (0.270)	−5.358** (0.266)	−4.188** (0.220)	−4.124** (0.219)
Male	1.576** (0.194)	1.587** (0.190)	2.476** (0.228)	2.385** (0.226)
AFQT less than Cat IIIB or missing	2.873 (1.805)	3.015† (1.780)	−0.189 (1.714)	−0.403 (1.693)
AFQT Cat IIIB	0.929** (0.211)	0.892** (0.208)	0.459+ (0.238)	0.402+ (0.236)
AFQT Cat II	−1.204** (0.191)	−1.121** (0.188)	0.180 (0.207)	0.110 (0.205)
AFQT Cat I	−3.068** (0.383)	−2.808** (0.378)	−0.458 (0.448)	−0.370 (0.444)
White	−0.831** (0.229)	−0.802** (0.225)	−0.250 (0.263)	−0.210 (0.261)
Black	2.468** (0.269)	2.401** (0.265)	0.475 (0.304)	0.493 (0.302)
Promoted rapidly	−0.917** (0.251)	−1.125** (0.247)	5.231** (0.375)	4.950** (0.373)
No. of observations	87,707	86,173	42,470	42,161

NOTES: Cell entries are estimated Tobit regression coefficients. Models also include controls for service duty MOS fixed effects and year-of-decision indicators. Standard errors are in parentheses.
** Denotes statistical significance at the 1 percent level.
* Denotes statistical significance at the 5 percent level.
† Denotes statistical significance at the 10 percent level.

Average SRBM, by Occupation, All Services

Table E.1
Average SRBM, by Two-Digit MOS, Army, First Term, FY 2002–2007

MOS	2002	2003	2004	2005	2006	2007
11	1.32	0.24	0.24	1.38	1.81	1.45
12	0.21	0.22				
13	0.94	0.20	0.18	0.85	1.50	1.26
14	1.20	0.23	0.13	0.81	1.62	1.26
15		0.32	0.29	0.95	1.27	1.49
19	0.84	0.21	0.23	1.10	1.37	1.17
21	0.65	0.32	0.21	0.80	1.57	1.15
25	0.05	0.02	0.26	0.94	1.34	1.32
31	1.06	0.18	0.15	1.67	2.23	1.52
35	0.57	0.10	0.17	1.03	1.51	0.52
42	0.00	0.06	0.09	0.52	0.72	0.90
44	0.07	0.03	0.09	0.71	1.17	0.76
52	0.08	0.07	0.21	1.05	1.63	0.95
54	1.01	0.41				
62	0.10	0.12	0.19	1.71	1.96	0.99
63	0.21	0.09	0.19	1.48	1.78	1.37
67	1.00	0.47				
68	0.90	0.47		1.72	1.78	1.23
71	0.12	0.03	0.00			
74	1.16	0.33	0.30	1.29	2.08	1.36
75	0.03	0.02		0.00		
77	0.79	0.19	0.00			
88	0.09	0.04	0.18	1.63	2.03	1.62
89		0.00	0.53	1.75	2.56	2.25
91	0.34	0.13	0.17	1.34	1.98	1.78
92	0.41	0.12	0.16	0.91	1.46	1.12
94				1.38	1.86	1.64
95	1.00	0.34				
96	1.40	0.19	0.24	1.62	2.26	2.08
98	2.27	1.30	0.61	0.95	1.35	1.53

NOTES: Data for 2007 cover January–September. Results are for occupations with at least 150 first- and second-term individuals in at least one year of the data from FY 2002 to FY 2007.

Table E.2
Average SRBM, by Two-Digit MOS, Navy, First Term, FY 2002–2007

MOS	2002	2003	2004	2005	2006	2007
AB	0.70	0.11	0.08	1.34	1.10	0.86
AC	1.79	0.28	0.84	3.06	2.57	2.44
AD	0.67	0.32	0.28	1.01	0.73	0.43
AE	2.71	2.15	2.23	1.03	0.36	0.05
AK	0.02					
AM	3.18	2.52	2.13	1.34	0.82	0.07
AO	3.04	3.60	3.50	2.27	1.39	1.48
AS	0.35	0.59	0.52	0.10	0.00	0.01
AT	2.60	1.45	1.30	0.66	0.40	0.99
AW	2.16	2.74	3.07	3.31	1.42	2.51
AZ	0.04	0.00	0.01	0.01	0.01	0.00
BM	0.08	0.06	0.08	1.29	1.14	0.82
CS			1.67	2.20	1.62	0.47
CT	3.32	2.06	1.55	1.47	1.47	1.67
DC	0.20	0.21	0.18	1.48	1.12	0.91
EM	2.98	3.14	2.66	3.34	3.42	3.35
EN	0.56	0.43	0.29	1.57	1.19	0.89
ET	4.98	5.68	5.98	5.65	4.88	4.13
FC	5.01	3.95	3.57	4.37	5.55	5.82
GM	0.67	0.89	1.08	1.37	0.50	0.26
GS	2.71	2.02	1.28	0.31	0.83	1.57
HM	1.15	0.77	0.50	1.26	1.43	1.23
HT	2.23	0.73	0.73	1.17	1.63	1.06
IC	1.35	0.38	0.62	1.31	0.75	0.44
IS	2.52	0.88	1.36	4.10	3.63	3.98
IT	2.69	2.53	2.65	3.11	2.08	1.84
MA	1.31	1.28	1.70	2.16	1.27	0.42
MM	3.80	4.00	3.25	3.27	3.01	2.65
MS	1.49	1.80				
OS	1.43	1.05	1.17	1.15	0.89	0.75
PN	0.04	0.12	0.00	0.02		
PS				0.02	0.00	0.02
QM	1.34	0.36	0.19	2.07	1.49	1.00
SH	0.63	0.24	0.11	0.42	0.22	0.00
SK	0.30	0.45	0.52	0.58	0.38	0.49
ST	4.30	3.87	3.65	4.16	3.94	4.09
YN	0.35	0.23	0.28	0.46	0.30	0.09

NOTES: Data for 2007 cover January–September. Results are for occupations with at least 150 first- and second-term individuals in at least one year of the data from FY 2002 to FY 2007.

Table E.3
Average SRBM, by Two-Digit MOS, Marine Corps, First Term,
FY 2002–2007

MOS	2002	2003	2004	2005	2006	2007
	0.49	0.00	0.00	0.00		1.97
01	0.03	0.04	0.00	0.02	0.02	0.77
02	1.07	1.87	3.85	4.07	3.05	5.03
03	0.79	0.77	1.77	3.26	3.06	4.37
04	0.93	0.85	0.80	0.75	0.93	2.34
06	0.97	0.82	0.69	1.02	0.92	2.30
13	0.00	0.01	0.00	0.01	0.36	2.17
28	2.46	1.98	1.68	1.76	1.70	4.18
30	0.00	0.00	0.00	0.06	0.01	0.80
35	0.00	0.00	0.01	0.00	0.30	2.22
60	0.86	0.52	0.10	0.57	0.82	2.11
61	0.69	0.86	0.86	1.30	1.75	3.22
65	0.91	0.95	0.97	0.87	1.14	2.61
80						2.47
99	0.49	0.65	0.34	0.72	0.98	1.90
ZZ	0.37	0.59	0.73	1.03	1.04	

NOTES: Data for 2007 cover January–September. Results are for
occupations with at least 150 first- and second-term individuals in at
least one year of the data from FY 2002 to FY 2007.

Table E.4
Average SRBM, by Two-Digit MOS, Air Force, First Term,
FY 2002–2007

MOS	2002	2003	2004	2005	2006	2007
1A	3.09	3.79	3.64	3.51	3.28	2.91
1C	4.13	4.25	3.59	2.64	2.76	1.87
1N	4.27	4.19	4.22	4.11	4.19	4.01
2A	2.99	3.21	2.35	0.97	0.32	0.29
2E	2.94	3.00	2.47	1.05	0.50	0.19
2F	2.38	2.62	1.84	0.49	0.44	0.44
2S	0.62	0.57	0.17	0.00	0.01	0.02
2T	1.50	1.35	1.02	0.40	0.44	0.47
2W	2.84	3.00	1.88	0.31	0.35	0.28
3A	0.22	0.18	0.15	0.01	0.00	0.00
3C	5.62	5.31	4.09	0.83	0.62	0.38
3E	2.84	3.17	2.40	0.79	0.40	0.30
3M	0.62	0.59	0.26	0.02	0.01	0.05
3P	3.42	3.65	2.58	1.13	0.42	0.46
3S	0.15	0.19	0.07	0.06	0.01	0.00
4A	0.88	1.04	0.44	0.03	0.00	0.02
4N	1.73	1.74	0.76	0.08	0.09	0.07
4Y	1.11	1.07	0.54	0.01	0.02	0.02
6F	0.71	1.06	0.13	0.02	0.00	0.01

NOTES: Data for 2007 cover January–September. Results are for
occupations with at least 150 first- and second-term individuals in at
least one year of the data from FY 2002 to FY 2007.

Table E.5
Average SRBM, by Two-Digit MOS, Army, Second Term,
FY 2002–2007

	2002	2003	2004	2005	2006	2007
11	1.16	0.68	0.72	1.79	2.33	1.80
12	0.43	0.44	0.50			
13	0.89	0.38	0.61	1.37	1.98	1.51
14	1.62	1.04	0.90	1.40	2.06	1.60
15		0.48	0.57	1.19	1.52	1.72
19	0.85	0.28	0.52	1.29	1.68	1.58
21		0.27	0.37	0.94	1.80	1.39
25	0.10	0.21	0.58	1.31	1.70	1.71
31	1.08	0.47	0.47	1.64	2.35	1.83
35	0.81	0.39	0.69	0.97	2.01	1.27
42	0.00	0.07	0.10	0.48	0.87	1.09
44	0.30	0.19	0.36	0.77	1.01	1.12
52	0.30	0.44	0.27	0.92	1.53	1.32
54	0.95	0.74			2.52	
62	0.28	0.30	0.29	1.36	2.15	1.24
63	0.40	0.29	0.32	1.33	1.92	1.63
67	0.96	0.60				
68	0.84	0.49	0.00	1.56	1.56	1.34
71	0.09	0.02	0.00	0.00		
74	1.10	0.52	0.71	2.00	2.63	1.64
75	0.08	0.04	0.50			
77	0.63	0.22	0.50			
88	0.16	0.12	0.19	1.41	1.96	1.68
89		1.48	0.76	1.91	2.85	2.42
91	0.28	0.19	0.26	1.04	1.68	2.03
92	0.40	0.14	0.27	0.95	1.44	1.34
94				1.18	1.76	1.65
95	1.07	0.44				
96	1.80	0.90	1.01	2.25	2.80	2.62
98	2.75	1.76	1.81	2.31	2.53	2.03

NOTES: Data for 2007 cover January–September. Results are for
occupations with at least 150 first- and second-term individuals in at
least one year of the data from FY 2002 to FY 2007.

Table E.6
Average SRBM, by Two-Digit MOS, Navy, Second Term,
FY 2002–2007

MOS	2002	2003	2004	2005	2006	2007
AB	0.22	0.27	0.54	0.81	1.05	0.94
AC	3.07	2.80	2.70	3.21	2.42	2.03
AD	0.96	0.76	0.98	1.23	0.77	0.64
AE	2.57	2.71	2.96	2.34	1.93	1.63
AK	0.06	0.00				
AM	2.03	2.05	2.28	2.35	2.41	1.84
AO	1.51	1.72	1.97	2.49	2.38	2.43
AS	0.34	0.21	0.23	0.28	0.15	0.26
AT	3.00	2.71	2.68	2.70	2.51	1.83
AW	2.46	2.27	2.89	2.72	2.40	2.22
AZ	0.05	0.05	0.00	0.00	0.01	0.01
BM	0.18	0.18	0.27	0.82	0.50	0.23
CS			1.11	1.49	1.53	1.29
CT	3.11	2.41	2.54	2.42	2.14	1.99
DC	0.39	0.29	0.43	1.04	0.43	0.15
EM	2.47	3.29	3.57	3.34	3.42	3.73
EN	0.69	0.63	0.38	1.04	0.76	0.48
ET	4.48	4.63	4.22	4.19	4.20	3.73
FC	4.10	3.11	3.03	3.60	3.83	3.75
GM	0.81	0.89	1.05	1.09	0.84	0.51
GS	2.72	2.31	1.31	1.04	1.44	1.46
HM	0.99	0.97	1.09	1.62	1.51	1.71
HT	1.40	0.76	1.12	1.22	1.31	0.66
IC	0.90	0.70	0.61	1.01	0.72	0.56
IS	3.29	2.11	2.35	3.47	3.73	3.26
IT	2.93	2.87	3.12	3.27	2.73	2.28
MA	0.86	1.66	2.05	2.17	2.51	2.77
MM	3.67	3.98	4.19	4.24	3.54	3.27
MS	0.79	0.96	3.02			
OS	1.68	1.70	1.81	1.69	1.61	1.09
PN	0.11	0.06	0.13	0.13		
PS				0.08	0.06	0.02
QM	1.51	2.19	1.20	1.89	1.61	1.04
SH	0.59	0.49	0.30	0.47	0.25	0.09
SK	0.33	0.29	0.48	0.95	0.55	0.39
ST	4.12	4.13	4.85	4.99	4.43	3.75
YN	0.19	0.20	0.21	0.24	0.25	0.21

NOTES: Data for 2007 cover January–September. Results are for
occupations with at least 150 first- and second-term individuals in at
least one year of the data from FY 2002 to FY 2007.

Table E.7
Average SRBM, by Two-Digit MOS, Marine Corps, Second Term,
FY 2002–2007

MOS	2002	2003	2004	2005	2006	2007
	0.40	2.96	0.00	0.00	0.00	0.97
01	0.01	0.01	0.00	0.00	0.01	0.49
02	1.07	1.62	1.68	2.35	2.94	4.44
03	0.39	0.48	0.48	0.91	1.05	2.09
04	0.20	0.15	0.08	0.20	0.49	1.18
06	0.20	0.34	0.40	0.40	0.64	1.43
13	0.03	0.10	0.00	0.00	0.24	1.09
28	1.98	2.05	1.76	1.24	1.30	2.37
30	0.00	0.01	0.02	0.01	0.04	0.53
35	0.00	0.02	0.00	0.04	0.07	0.89
60	0.42	0.38	0.14	0.05	0.32	0.96
61	0.97	1.53	1.22	0.91	1.08	1.32
65	0.09	0.08	0.01	0.41	0.57	1.57
80						1.31
99	0.28	0.25	0.15	0.18	0.28	0.64
ZZ	0.31	0.40	0.20	0.29	0.47	

NOTES: Data for 2007 cover January–September. Results are for
occupations with at least 150 first- and second-term individuals in at
least one year of the data from FY 2002 to FY 2007.

Table E.8
Average SRBM, by Two-Digit MOS, Air Force, Second Term,
FY 2002–2007

MOS	2002	2003	2004	2005	2006	2007
1A	2.71	3.15	3.03	2.64	2.84	3.05
1C	3.81	4.75	4.29	3.31	3.79	3.42
1N	4.39	5.05	5.87	5.17	5.60	4.57
2A	1.72	2.40	2.40	1.63	1.20	1.09
2E	2.41	2.74	2.93	1.75	1.69	1.20
2F	0.33	0.78	0.84	0.67	0.66	0.75
2S	0.45	0.40	0.07	0.07	0.23	0.22
2T	0.53	0.66	0.86	0.79	0.74	0.71
2W	1.68	2.24	1.93	0.99	1.09	1.00
3A	0.45	0.38	0.06	0.06	0.08	0.06
3C	5.51	6.35	5.58	2.87	2.53	2.41
3E	1.53	2.17	2.04	1.33	1.41	1.20
3M	0.84	0.79	0.22	0.37	0.43	0.33
3P	1.49	1.88	1.98	1.68	1.46	1.48
3S	0.74	0.71	0.12	0.23	0.14	0.17
4A	0.56	0.46	0.09	0.11	0.35	0.46
4N	0.43	0.48	0.65	0.63	0.66	0.75
4Y	0.93	1.17	1.07	0.55	0.61	0.55
6F	0.52	0.63	0.43	0.36	0.38	0.31

NOTES: Data for 2007 cover January–September. Results are for
occupations with at least 150 first- and second-term individuals in at
least one year of the data from FY 2002 to FY 2007.

Distribution of Bonuses

Table F.1
Distribution of SRBM, by Service, FY 2007

SRBM	Bonus Not Conditional on Deployment				Bonus Conditional on Deployment			
	Army	Navy	Marine Corps	Air Force	Army	Navy	Marine Corps	Air Force
First Term								
0	0.14	0.21	0.09	0.59	0.06	0.24	0.10	0.69
0.5	0.09	0.16	0.07	0.06	0.04	0.17	0.07	0.06
1	0.10	0.26	0.10	0.06	0.12	0.26	0.10	0.06
1.5	0.26	0.09	0.09	0.04	0.30	0.10	0.09	0.04
2	0.23	0.06	0.04	0.04	0.34	0.06	0.06	0.04
2.5	0.07	0.05	0.07	0.01	0.09	0.06	0.06	0.02
3	0.03	0.02	0.07	0.01	0.03	0.02	0.07	0.01
>3	0.07	0.14	0.47	0.18	0.03	0.09	0.45	0.07
Second Term								
0	0.12	0.17	0.20	0.20	0.08	0.23	0.22	0.21
0.5	0.05	0.13	0.22	0.17	0.04	0.13	0.23	0.18
1	0.10	0.13	0.14	0.17	0.12	0.12	0.14	0.18
1.5	0.19	0.07	0.09	0.16	0.20	0.08	0.09	0.16
2	0.24	0.08	0.05	0.09	0.27	0.08	0.05	0.09
2.5	0.16	0.07	0.08	0.05	0.16	0.07	0.08	0.06
3	0.07	0.05	0.07	0.03	0.08	0.05	0.07	0.03
>3	0.08	0.30	0.16	0.13	0.06	0.24	0.12	0.11

Bibliography

Asch, Beth, James Hosek, and John Warner, "New Economics of Manpower in the Post–Cold War Era," in Todd Sandler and Keith Hartley, eds., *Handbook of Defense Economics,* Vol. 2, Amsterdam: Elsevier, 2007, pp. 1076–1138.

Asch, Beth J., James R. Hosek, Michael G. Mattock, and Christina Panis, *Assessing Compensation Reform: Research in Support of the 10th Quadrennial Review of Military Compensation,* Santa Monica, Calif.: RAND Corporation, MG-764-OSD, 2008. As of February 22, 2010:
http://www.rand.org/pubs/monographs/MG764/

Asch, Beth J., Paul Heaton, and Bogdan Savych, *Recruiting Minorities: What Explains Recent Trends in the Army and Navy?* Santa Monica, Calif.: RAND Corporation, MG-861-OSD, 2009. As of February 22, 2010:
http://www.rand.org/pubs/monographs/MG861/

Ash, Colin, Bernard Udis, and Robert McNown, "Enlistments in the All-Volunteer Force: A Military Personnel Supply Model and Its Forecasts," *American Economic Review,* Vol. 73, 1983, pp. 144–155.

Cameron, A. Colin, and Pravin K. Trivedi, *Microeconometrics: Methods and Applications,* Cambridge, U.K.: Cambridge University Press, 2005.

College Board, *Trends in College Pricing,* Washington, D.C., 1998–2008.

Department of the Army, *Fiscal Year Budget Estimates: Military Personnel Army Justification Book,* various years. As of March 4, 2010:
http://www.asafm.army.mil/budget/fybm/fybm.asp

Department of Defense, *Defense Budget Materials,* various years. As of September 23, 2009:
http://www.defenselink.mil/comptroller/defbudget/fy2009/index.html

———, *Report of the 9th Quadrennial Review of Military Compensation,* Office of the Under Secretary of Defense for Personnel and Readiness, Washington, D.C., 2002.

———, *Military Compensation Background Papers*, Washington, D.C., 2005.

———, *The Military Compensation System: Completing the Transition to an All-Volunteer Force,* Report of Defense Advisory Committee on Military Compensation, Washington, D.C.: Office of the Under Secretary of Defense for Personnel and Readiness, 2006. As of September 14, 2009:
http://www.defenselink.mil/prhome/dacmc.html

———, *Military Recruiting Results,* 2009. As of September 23, 2009:
http://www.defenselink.mil/prhome/mpprecruiting.html

———, *Financial Management Regulation 7000.14-R*, 2010. As of March 25, 2010:
http://comptroller.defense.gov/fmr/07a/index.html

Dertouzos, James N., *Recruiter Incentives and Enlistment Supply,* Santa Monica, Calif.: RAND Corporation, R-3065-MIL, 1985. As of February 22, 2010:
http://www.rand.org/pubs/reports/R3065/

Dertouzos, James N., *The Cost-Effectiveness of Military Advertising: Evidence from 2002–2004,* Santa Monica, Calif.: RAND Corporation, DB-565-OSD, 2009. As of February 22, 2010:
http://www.rand.org/pubs/documented_briefings/DB565/

Dertouzos, James N., and Steven Garber, *Is Military Advertising Effective? An Estimation Methodology and Applications to Recruiting in the 1980s and 90s,* Santa Monica, Calif.: RAND Corporation, MR-1591-OSD, 2003. As of March 19, 2009:
http://www.rand.org/pubs/monograph_reports/MR1591/

Goldberg, Matthew S., *A Survey of Enlisted Retention: Models and Findings,* Alexandria, Va.: Center for Naval Analyses, CRM D0004085.A2/Final, 2001.

Goldberg, Matthew, and John Warner, *Determinants of Navy Reenlistment and Extension Rates,* Alexandria, Va.: Center for Naval Analyses, Research Contribution 476, 1982.

Hansen, Michael L., and Jennie W. Wenger, *Why Do Pay Elasticity Estimates Differ?* Alexandria, Va.: Center for Naval Analyses, CRM D0005644.A2/Final, November 2002.

———, "Is the Pay Responsiveness of Enlisted Personnel Decreasing?" *Defence and Peace Economics,* Vol. 16, No. 1, 2005, pp. 29–43.

Hattiangadi, Anita U., Deena Ackerman, Theresa H. Kimble, and Aline O. Quester, *Cost-Benefit Analysis of Lump Sum Bonuses for Zone A, Zone B, and Zone C Reenlistments: Final Report,* Alexandria, Va.: Center for Naval Analyses, CRM D0009652.A4/1Rev, May 2004.

Heckman, James, and Jeffrey Smith, "Assessing the Case for Social Experiments," *Journal of Economic Perspectives,* Vol. 9, No. 2, 1995, pp. 85–110.

Heckman, James, and Salvador Navarro-Lozano, "Using Matching, Instrumental Variables, and Control Functions to Estimate Economic Choice Models," *Review of Economics and Statistics,* Vol. 86, No. 1, 2004, pp. 30–57.

Hogan, Paul, and Brian Simonson, *Navy SRB Effect on the Length of Reenlistment,* Falls Church, Va.: The Lewin Group, 2007.

Hogan, Paul, T. Dali, Patrick Mackin, and C. Mackie, *An Econometric Analysis of Navy Television Advertising Effectiveness,* Falls Church, Va.: Systems Analytic Group, 1996.

Hogan, Paul, Curtis Simon, and John Warner, "Sustaining the Force in an Era of Transformation," in Barbara Bicksler, Curtis Gilroy, and John Warner, eds., *The All-Volunteer Force: Thirty Years of Service,* Washington, D.C.: Brasseys, Inc., 2004, pp. 57–89.

Hogan, Paul, Javier Espinosa, Patrick Mackin, and Peter Greenston, *A Model of Army Reenlistment Behavior: Estimates of the Effects of the Army's Selective Reenlistment Bonus on Retention by Occupation,* Arlington, Va.: U.S. Army Research Institute for the Behavioral and Social Sciences, December 2005.

Hosek, James R., and Mark Totten, *Serving Away from Home: How Deployments Influence Reenlistment,* Santa Monica, Calif.: RAND Corporation, MR-1594-OSD, 2002. As of March 9, 2010:
http://www.rand.org/pubs/monograph_reports/MR1594/index.html

Hosek, James R., and Francisco Martorell, *How Have Deployments During the War on Terrorism Affected Reenlistment?* Santa Monica, Calif.: RAND Corporation, MG-873-OSD, 2009. As of March 4, 2010:
http://www.rand.org/monographs/MG873/

McFadden, David, "Econometric Analysis of Qualitative Response Models," in Zvi Griliches and Michael Intrilligator, eds., *Handbook of Econometrics,* Vol. 2, Amsterdam: Elsevier, 1983.

Milpers Memorandum Number 08-324, December 19, 2008.

Milpers Message Number 04-355, Army Human Resources Command, December 30, 2004.

Milpers Message Number 06-007, Army Human Resources Command, January 5, 2006.

Milpers Message Number 07-141, Army Human Resources Command, June 6, 2007.

Milpers Message Number 07-344, Army Human Resources Command, December 12, 2007.

Milpers Message Number 08-068, Army Human Resources Command, March 13, 2008.

Moffitt, Robert, "The Role of Randomized Field Trials in Social Science Research: A Perspective from Evaluations of Reforms of Social Welfare Programs," *American Behavioral Scientist,* Vol. 47, No. 5, 2004, pp. 506–540.

Moore, Carol, Paul Hogan, Kristen Kirchner, Patrick Mackin, and Peter Greenston, *Econometric Estimates of Army Retention: Zones A, B, C, D and Retirement-Eligible Estimates with Data Through FY 2004,* Arlington, Va.: U.S. Army Research Institute for the Behavioral and Social Sciences, November 2006.

Plümper, Thomas, and Vera E. Troeger, "Efficient Estimation of Time-Invariant and Rarely Changing Variables in Finite Sample Panel Analyses with Unit Fixed Effects," *Political Analysis,* Vol. 15, No. 2, 2007, pp. 124–139.

Polich, J. Michael, James N. Dertouzos, and S. James Press, *The Enlistment Bonus Experiment,* Santa Monica, Calif: The RAND Corporation, R-3353-FMP, 1986. As of March 4, 2010: http://www.rand.org/reports/R3353/

Simon, Curtis J., and John T. Warner, "Managing the All-Volunteer Force in a Time of War," *Economics of Peace and Security Journal,* Vol. 2, No. 1, 2007, pp. 20–29.

———, "The Supply Price of Commitment: Evidence from the Air Force Enlistment Bonus Program," *Defence and Peace Economics,* Vol. 20, No. 4, 2009, pp. 269–286.

Simon, Curtis J., Sebastian Negrusa, and John T. Warner, "Educational Benefits and Military Service: An Analysis of Enlistment, Reenlistment, and Veterans' Benefit Usage 1991–2005," *Economic Inquiry,* forthcoming, 2010.

Tsui, Flora, Paul Hogan, Jeff Chandler, Javier Espinosa, Patrick Mackin, and Peter Greenston, *Army SRB Management Model: Econometric Estimates of Effects on Retention and Length of Reenlistment,* Arlington, Va.: U.S. Army Research Institute for the Behavioral and Social Sciences, March 2005.

Waller, Douglas, "This Man Wants You," *Time,* February 9, 2003. As of March 4, 2010: http://www.time.com/time/magazine/article/0,9171,421044,00.html

Warner, John T., and Beth J. Asch, "The Economics of Military Manpower," in K. Hartley and T. Sandler, eds., *Handbook of Defense Economics,* Vol. 1, Amsterdam: Elsevier, 1995, pp. 347–398.

Warner, John, Curtis Simon, and Deborah Payne, *Enlistment Supply in the 1990s: A Study of the Navy College Fund and Other Enlistment Incentive Programs,* Washington, D.C.: Defense Manpower Data Center, DMDC Report No. 2000-015, 2001.

Wooldridge, Jeffrey, *Econometric Analysis of Cross Section and Panel Data,* Boston, Mass.: MIT Press, 2001.